ROYAL JERSEY AGRICULTURAL & HORTICULTURAL SOCIETY
EST 1833

Jersey's Rural Heritage

A farming way of life

Written and compiled by Derrick Frigot

Edited by James W. Godfrey MSc FRAgS MRAC

"Beautiful Jersey"

Man bieau p'tit Jèrri, la reine des îles
Lieu dé ma naissance, tu m'pâsses près du tchœu
Ô, tchi doux souv'nîn du bouôn temps qu' j'ai ieu,
Quand j'pense à Jèrri, la reine des îles.

My beautiful little Jersey, queen of the isles
Place of my birth, close to my heart
O, sweet memories of the good times I've had
When I think of Jersey, queen of the isles.

Published by the

Royal Jersey Agricultural & Horticultural Society
2001
ISBN 0-9452011-0-8
In conjunction with Partners in Print, Jersey
Printed and bound in Great Britain by Butler & Tanner Ltd, Frome and London

the Jersey Royal

Jersey's Rural Heritage

omatoes and Flowers

the Jersey Cow

the Harvest

the R.J.A. & H.S.

a world of Jerseys

contents

ROYAL · JERSEY · AGRICULTURAL & HORTICULTURAL · SOCIETY

EST 1833

foreword

The beautiful landscape of Jersey has been shaped for generations by the activities of man through agriculture, in a relationship which touches all our lives.The story of Jersey's agricultural industry is a remarkable one of which we should be proud.

Farming is a long-term business, undertaken with patient determination, working within the natural cycles of the seasons. It provides a thread of continuity that runs through the generations, whether in farming families, cattle breeding or the husbandry of the land.

There have been many changes to the rural scenery over the years and this process will undoubtedly continue. We must cherish our rural heritage, nurture our countryside and allow our industry to flourish in the future.

This book is a reflection of a farming way of life through the centuries, the last two under the leadership and guidance of the Royal Jersey Agricultural & Horticultural Society. I sincerely hope that this book gives you many happy hours of enjoyment.

Lewis J Rondel
President
Royal Jersey Agricultural & Horticultural Society
November 2001

acknowledgements

The Royal Jersey Agricultural & Horticultural Society has been in existence for 168 years and it is natural that an organisation of this kind, which has had an influential role in Island affairs and links with cattle breeders and organisations in so many countries all around the world, should have an extensive collection of memorabilia. It was during the move into the new headquarters at Trinity that the idea was conceived to pool our photographic resources and publish them in a book.

In an attempt to provide a more complete photographic record of Jersey farming through the ages, we asked members and friends of the Society to submit family, or other pictures for consideration. The response was excellent and many pictures in this book come from private collections.

We are very appreciative of the extensive help given to us by the Jersey Evening Post, and their photographic department, in particular, archivist, Jan Hadley.
The JEP has a wonderful photographic record of Island life during the last century and they were extremely kind in allowing us access to their archives.

In addition, the Société Jersiaise was helpful in offering pictures from their fine collection that allowed us to use some fascinating historical pictures of farming life in Jersey.
Robin Briault was always very helpful and obliging in supplying high quality prints from the many negatives we used.

A special thanks is due to Peter Lee for so conscientiously reading through the original text and correcting where necessary and also adding so much of his wide knowledge of farming. Also to Winston Le Brun whose knowledge of farm life in Jersey "temps passé" and the Jersey patois greatly enhanced the captions to the pictures.

James Godfrey, the Society's Chief Executive, has played a vital role in the production of the book. His contribution of ideas along with his regimental approach to its production has been invaluable. The members of the Society's staff, Ann Guillard, David Hambrook, Pam Laurens and Gwynneth Morgan have all been very helpful in the preparation of the book.

The artistic ability of Eddie McGrath has been a key factor in the design and production of the book. We are grateful for his time and patience.

Grateful thanks also to Chris Minty for his help with the printing of this book.

In addition we would like to thank the following for all their help, in most cases for the loan of pictures and other memorabilia for this book: Stuart Abraham, Ron and Jack A'court, Tommy A'Court, the late Arthur Ahier, Brian Ahier, Jean Arthur, Harry Baudains, John and Margaret Baudains, Collette Bisson, the late Joe Boleat, Catherine Le Breton, Philip Le Brocq, Frederick Brown, Basil Le Brun, Ernest Le Conte, Charles Le Cornu, Marjorie Dodds, Frank Falle, Nan du Feu, Steve and Suzanne Le Feuvre, John Le Gallais, Sue Hardy, Graham Harris, Vicky Huelin, Jersey Archive Centre, Jersey Milk, Pam Laurens, The Lempriere Collection, Hedley Maillard, Stuart McAlister, Jennifer McMullon, Paul Le Miere, Luke Le Moignan, Ian Monins, National Trust for Jersey, Anne Perchard, Les Quenault, Reg Queree, Nick Le Quesne, Margaret Renouf, Maurice Richard, Denis Richard, Maurice Richardson, Roger Rive, Daphne Romeril, John Rondel, Richard Rondel, Maria Ruellan, Joyce Le Ruez, Les and Mazel Le Ruez, Don Le Sech, Oswald Simon, Sheila Staite, Jack and Eileen Le Sueur, Justin Surcouf, Ted Syvret, Michael and Nancy Vautier, Christine Vibert, Lis Vivian, John Vint, Aldyth Walker, Norman Walker, Vicky Warren.

Writing and compiling *Jersey's Rural Heritage – a farming way of life* has been absorbing and full of interest. Apart from reviving long lost memories of childhood days, it has given me the opportunity to renew old friendships, and to meet new acquaintances. This has been a very special time and I thank everyone for their input in whatever way it was given.

Derrick Frigot
November 2001

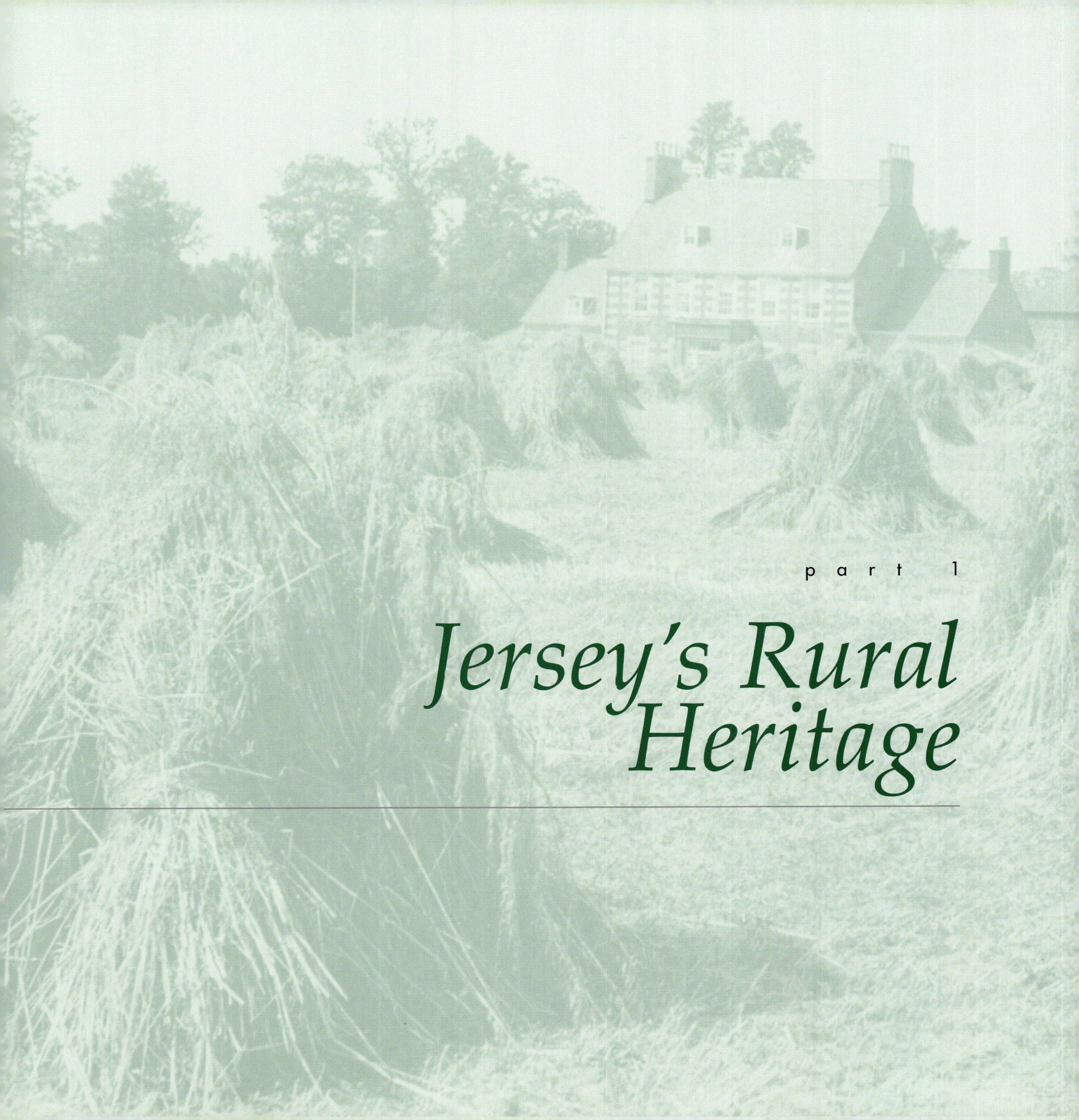

part 1

Jersey's Rural Heritage

Jersey's Rural Heritage

Jersey is favoured for agriculture with its south-facing aspect that warms the soil early in the spring. Its light, loamy texture dries out quickly making it easy to work, and the surrounding sea keeps the island almost frost-free in the wintertime.

The Island's countryside, as we know it today, has been fashioned through centuries of changing agricultural history. Five hundred years ago, the rural view of Jersey was vastly different from today with fields of corn, much common land and few trees or hedges.

Jerseymen have always been an enterprising and ambitious race. As an island, there has always been strong links with the sea, and in the days before steamships, shipbuilding was an important industry in the Island. At one time the Jersey fleet had as many as 500 vessels. Jerseymen were some of the first to fish for cod off the Newfoundland bank and they sailed in their small boats across the Atlantic every spring, returning home in time for the autumn ploughing. Having caught and salted the cod, they moved on to the ports of Spain, the Mediterranean and even to North and South America, trading their cargo that included knitted garments from the Island, for other valuable goods. In times of war, they took on the role of privateers for the English Navy, which, as a form of 'licensed pirate' gave them authority to plunder enemy ships. Reputedly, they were also not averse to smuggling, and these

• A 19th century farmyard.

• St Peter's Valley in the 19th century.

Land in Jersey is traditionally measured in 'vergées'.
One acre is equal to two and quarter vergées.

- An old thatched cottage and farm buildings dating from about 1500 A.D. (c.1919)

- Mending nets at Gorey (c.1938)

- Bringing in the cows for milking.

Jersey's Rural Heritage

various enterprises, both legal and illegal, helped to amass great wealth for the Island.

With this strong maritime tradition, the Island was in a good position to export homegrown products. Taking full advantage of the situation, farmers turned from growing corn, mostly wheat that did little to enhance their income, to keeping sheep in great numbers and producing and exporting fine knitted clothing.

At that time, Jersey's native sheep were depicted as being peculiar to the Island. They were a small breed, and were described by Jean Poingdestre, the Jersey historian, in 1682, as "the females had most foure hornes and the rams oft times six, that is three on each side whereof two made a circle towards the nose, two others another circle backwards towards the ears and two stood upright between them". He also reported at that time that these native sheep were almost extinct and were being replaced by a larger breed, "like those in Salisbury plain".

Even though there were many flocks in Jersey - including more than 300 flocks in the parish of St Ouen alone – the knitting industry was so large that additional wool was imported from England. However, an Act of Parliament restricted this in 1660 to 4,000 tods, an old English wool measure equivalent to about

• A 1950s harvest scene.

• "Counting sheep in the 1950s!"

• An old thatched farm – the lady is drawing water from the well.

• Sheep were traditionally kept on the poorer coastal ground in the 1950s.

Jersey's Rural Heritage

twenty-eight pounds.

Historical writers have quoted that from 6,000 to 10,000 pairs of stockings were exported weekly, going to England, France and Spain. This trade was lucrative for the Islanders but there was a price to conventional farming. They neglected their crops, and in 1608, a law was passed forbidding every person over the age of 15 from knitting during the vraicing and corn harvesting seasons. The word "jersey" given to a knitted sweater derived its name from the Island's knitting industry that lasted for an astounding 250 years.

With the wealth that the maritime and knitting industries brought to the Island, farmers turned to other more rewarding enterprises. The cider industry became an important venture, complementing the work of knitting. This change of direction in farming, from corn to cider, encouraged the biggest transformation in the appearance of the Island. At one time, one quarter of Jersey was covered in apple orchards with tall hedges built up around them to provide shelter.

Between these high hedges, arches of trees cover winding narrow lanes. Part of the farmers work today stems from the construction of the hedges and byways – the annual branchage

- Cows in the orchard at J G Buesnel at St Saviour (c.1919)

- Overlooking St Ouen's Bay

- Chasing the heifers home, Roselands, St Saviour.

• Overlooking St. Brelade's Bay (c.1900).

• Young girl in traditional costume.

Jersey's Rural Heritage

which is the cutting of the trees and banks twice each year, in July and September.
The responsibility to oversee that all hedges are cut lies with the parochial authorities, but in August of each year the Royal Court conducts the Visite Royale in two parishes, inspecting the parish books, accounts, and the roads.

Several centuries ago, the farmer grew wheat and rye to make bread, and apples to make cider. Before the days of potatoes and other root crops, he also grew parsnips in abundance, to provide winter vegetables for his family, and for his livestock. The farm was self-sufficient, as, apart from the crops, the farmer had a cow or two for milk and butter, sheep for wool and mutton and pigs for pork and bacon. A few fowls provided eggs. Heating in the winter was generated by gorse, heather and vraic.

Jersey's soil was cared for by enriching it constantly with the manure from the livestock and the vraic from the seashore. The farmer rested his land between seasons, and ploughed in the cereal stubble, now covered with manure or vraic, making excellent use of the natural fertilizers. He also maintained permanent pasture, in the meadows and apple orchards, for his tethered cattle to graze.

• A typical Jersey farm scene in the 19th century.

• Visite de Branchage at St Peter in 1993.

• Branchage

• This little lad is certainly making friends with the chickens

• Corn stooks and heifers in 1945.

Jersey's Rural Heritage

It was probably the deep rooting parsnip that prompted the introduction of Le Grand' Tchéthue, the big plough, to the Island in 1768. The parsnip did indeed have a long root that required a good depth of cultivated soil to grow.

The gradual change from cider apples to potatoes occurred at the end of the 18th century, by which time there was a thriving and profitable cattle export business.

In the mid-19th century it was realised that the demand, and the money, lay in the production of early potatoes. This encouraged the development of Jersey's côtils; the south facing slopes of land near the island's coasts. Farmers grubbed up the gorse bushes, turning this previously unfarmed soil into the most profitable land in the Island.

Always searching for crops that would boost farm income, many island farmers started growing outdoor tomatoes on the lighter, sandy soils, in the early 1900s. Later, in the years following the German Occupation, significant fortunes were made from the tomato industry.

In the first half of the 20th century, Jersey's numerous small farms were still less than 50 vergées on average, but were, by and large, profitable small enterprises. With the exceptional periods of two World Wars and the great depression of the 1930s, Jersey's farmers made a

• Anne Port Farm (c.1900)

• Bouley Bay, showing the hillside terraced for potatoes (c.1919)

- Horse and van with barrels of parsnips, along Victoria Avenue. Driving the horse was George Baudains and riding with him was George Smith, grandfather of Basil Le Brun from Sion. (c.1910)

- Tall Jacks, or Jersey cabbages. These cabbages were grown primarily to feed livestock, particularly pigs, and grew to eight feet high, and more, providing leaves for fodder all the time.

Jersey's Rural Heritage

good living from the land. Farming progressed from the horse to the tractor and along with the traditional pattern of cattle and early potatoes, sometimes with tomatoes, came a variety of second crops like winter cauliflowers and calabrese, flower bulbs, and at times, varieties of flowers themselves. All had a part to play in the income for the farmer and his family.

The Jersey farmer has, for generations, faced challenges, usually from forces outside his influence, and these have not only affected his ability to farm, but have affected changes in the countryside. In the past it more often than not was a tactical change in crops prompted by competition from another source; today this is still the case, but the influence is far more global, both geographically and industrially.

Many of the granite farmhouses and extensive outbuildings are now fine residential properties. The surrounding fields are still farmed, but by farmers who have embraced the challenge of contemporary growing and marketing by expanding their land base, using modern technology. Still today, 53% of the island is worked and cared for by the farmers who strive, not only to make a living, but also to protect the environment.

- The gradual changeover from horse to tractor power. The reliable horse was used here to plant the potatoes while the tractor, fitted with cage wheels was drawing a guano spreader.

- Breezing, or shallow ploughing, near St Ouen's Mill (c.1970)

- Mrs Harold Slous from Corbiere, with one of her old favourites (c.1987)

- Coming home at Morel Farm, St Lawrence.

J33
173

Jersey's Rural Heritage

The unique and familiar countryside of Jersey, with its winding lanes, high banks with elm trees or hawthorn hedges planted on top, or perhaps solid granite walls, enclosing the small fields, dates from a change in agricultural practices some four hundred years ago.

Then, probably around the mid 16th century, smallholders started the process of converting their larger open fields of corn into apple orchards. The fruit trees needed shelter from the strong winds that whipped off the sea over the exposed land, and the farmers set about constructing these smaller enclosures to protect the new crop.

The cider industry expanded, with about one quarter of the land covered with apple trees. Like so many other activities that led to overproduction, the extensive planting of orchards had to be restricted by an act of the States in 1673, which forbade the planting of new orchards except to replace old ones.

Jersey, being located between Normandy to the east, where the cider trade was very significant, and the south of England where cider was beginning to become more and more popular, was in a favourable position. Cider produced in Jersey and shipped to England was exempt of duty, while the French had to pay duty on exports to their English neighbours.
At that time, delivery by sea was cheaper than transport overland, and it was probably easier and less costly to take the barrels of cider from Jersey to the mainland by ship and continue through England's inland waterways, than to transport Devon or

- "Springtime in Jersey" with tethered cows in orchards was a very typical scene in Jersey of yesteryear. This was Highstead Farm, St Saviour in May 1949.

- A typical country lane in Jersey. The high banks, planted with trees on top were bui t to protect the cider apple orchards.

Jersey's Rural Heritage

Somerset cider by horse and cart along the rutted tracks, which passed for roads in those days.

Jersey farmers grew the first apples to make cider for home consumption; producing a surplus of the amber liquid to sell to England was, along with knitting and selling of woollen stockings, a very profitable enterprise.

In 1692, the noted local historian, Reverend Phillippe Falle wrote…"I do not think there is any country in the world that, in the same extent of ground, produces so much cider as Jersey does, not even Normandy itself. Nor is there better, larger and more generous fruit that what grows in this Island." Obviously, the Reverend gentleman knew his cider!

Similar to later Acts and Laws of the States controlling the importation from France of cattle (to protect the local Jersey breed), acts prohibiting the importation of cider from Normandy were necessary. For the same reasons, there were always some who sought to profit from evading the payment of English duties on cider shipped from France.

The production of cider in Jersey reached considerable proportions in the halcyon days of the industry in the late 1700s and early 1800s when between one and a half and two million gallons were produced each year. About a million gallons was consumed locally, and the balance exported to England.

- "At peace" – Jersey's idyllic countryside.

- At John Dorey, the Cider Merchant, Brook Farm, St John (c. 1900)

- Another cider merchant – Mr A Le Riche at Trinity. Note the casks full of empty bottles ready to be filled for export.

- Crushing apples before they go on the press.

Jersey's Rural Heritage

But by the 1850s the export of Jersey cattle was more profitable to the farmer and cider exports dwindled, eventually dying out by the end of the century. When the sale of cider to England ceased, the orchards in Jersey were still producing their annual drop of fruit and for a while, large quantities of apples were sent to the new cider factories in England. For a time in the 1880s, vinegar, made from the cider, provided an additional export income for the Island.

Cider making not only changed the natural contours of Jersey's countryside, but also contributed to the style of granite farms, as we know them today. The profit from cider helped many farmers to invest considerable sums of money in the circular granite apple crushers with large granite wheels to crush the apples into pulp. They also installed cider presses and built granite outbuildings to house them. Much of the stone used for the crushers came from the French Channel Island of Chausey as it was softer and easier to work than the harder Jersey granite. The combination of apple orchards, crushers and cider press houses, led to the Island's countryside taking the shape we see today. Granite farmsteads, surrounded by small fields, set in a maze of lanes, Jersey's unique natural rural scene.

- Manor Farm,
  Vinchelez,
  St Ouen

- A beautiful Jersey lane at Badier Farm, St Lawrence.

- Cider-making

Jersey's Rural Heritage

Today, few Jersey cider makers still practise their craft, but in earlier times, most farmers processed the apples grown on their farms. Each year, from September to January, depending on the season, on crisp winter mornings, the farmer harnessed his horse to the big wheel, filled the circular trough with apples and began the process of crushing them into pulp.

The gigantic oak wooden press with its massive screwed shaft running through it was already prepared. The pulp was scooped out of the crusher and tipped onto large hessian sacks, or sheets – in earlier days straw was used – which were then folded over, on the press, "envelope style" ready for the next layer. This was known as "bagging up". When the layers of pulped apples filled the press, boards were placed on top for protection and the pressure began. The screw was turned gradually so that the juice was squeezed through the bottom into a large underground vat. The unfermented brew was ready for barrelling.

The juice was baled from the vat into barrels where it was left to stand and "boil" until ready for consumption. By the middle of the 20th century, cider production had declined considerably but was still a part of farm life. The amber liquid, the staple drink of all country folk for many generations, was, for the Breton workers who came to work for "the season" on the farms, a necessary part of life!

• The process of cider-making at George Boleat, the last commercial cider-maker in Jersey.
Crushing the apples into pulp, (above).

• Tipping the apples into the crusher.

• Scooping out the pulp.

• "Turning the screw" on the press.

Autumn brings the apples down from the trees, and this is the time when families and friends get together for the annual making of black butter. The occasions for "sethée d'nier beurre" – the black butter night.

The recipe for a batch of black butter resembles a shopping list to feed an army ... 20 gallons of rough, unfermented cider, 22 hundredweight of sweet apples, one hundredweight of cooking apples (Bramleys or similar) 22 lemons, eight sticks of liquorice, 28 pounds of white sugar, 1 pound of cinnamon, half a pound of nutmeg and one pound of mixed spices.

The first task is to set the fire, making sure there is plenty of wood in reserve for the hours ahead. Over the open fire a "trivet" or tripod is placed on which sits the "bachîn" or brass basin. The bachîn is greased with lard; the cider is poured in and left to simmer for a few hours. During this time, the womenfolk are busy peeling the apples and 28 pounds at a time are added to the bachîn. With the addition of the first lot of apples, the process of stirring starts, using a "rabot", a long-handle implement shaped like a large hoe.

The stirring, which must be continued throughout the long cooking time, gradually gets more and more difficult, as the mix thickens. It is important to keep stirring to prevent the butter from sticking to the sides of the bachîn and burning.

As each batch of apples dissolves into the mixture more sweet apples are added, then the liquorice goes in to give the rich deep colour, and of course,

- All pictures: Scenes at annual Black Butter events. Peeling apples requires all hands on deck. Black Butter nights were full of banter and fun.

that spicy taste. The Bramleys are the last of the fruit to go into the bachîn and then sugar is added. This thickens the mixture that by now takes two men to draw the rabot backwards and forwards in that regular motion.

The different spices are mixed in and with about an hour of the cooking time left, the grated rind and juices of the lemons are finally added to the recipe. Cooking the black butter over some twenty-four hours gives lots of time for the increasing group of helpers to indulge in what they have traditionally done since the ritual began; eat - usually a bean crock, drink and get merry on cider, with an accordionist to encourage singing and dancing through the night.
Next official task is the tasting of the mixture and when pronounced ready by the experts, four men, using hessian sacking as webbing, lift the heavy bachîn off the trivet and place it on the floor.

Hundreds of jam jars have been washed and the women start filling them with the hot butter. This is a noisy operation as each jar is banged down on the table to expel any trapped air; it would do no good to offer anyone less than a full jar of black butter! Finally, the cooling jars are covered with a paper seal, held on by an elastic band.

There is just one more thing to be done. The thick, dark brown butter is now ready to be spread on a piece of bread. Now that really is the delicious taste of old Jersey.

• The cooking is complete

Black Butter

• Stirring goes on for 24 hours

• Filling the jars with a black butter.

Jersey's Rural Heritage

Main Picture, middle:

'The Trinity Trotting Match', September 1938 – Farmers at play. Left to right: Bert Ahier, Helen Picot, Clarry Le Sueur, Horace Ahier

Smaller pictures, clockwise from top left:

1) Fresh Jersey cabbage loaf

2) Showing heavy horses at the Island Show, May 1938

3) The Jersey Drag Hunt passing Trinity Church, January 1938

4) Jack Vautier's winning entry in the Horse & Van Class, Island Summer Show 1955

5) Ormering at low tide, 1952

6) Sunday School outing from St. Lawrence

7) Edward Vautier in a patch of Jersey Cabbages

8) Typically decorated farm lorry for the annual St. Martin's Methodist Sunday School Treat

9) Wonder making

Rural Pastimes

part 2

Jersey Royal Potato

Jersey Royals

The King of Potatoes! Every January, the Island changes its cloak of dull grey winter with the planting of its most precious crop, the early potato. Not just any potato, but the celebrated Jersey Royal, which now enjoys EU protection of designation of origin … Jersey's own champagne image.

Ploughing the fields transforms Jersey into a new rich brown of tilled land, when the potato goes into the ground for its first tentative weeks. Today's calendar includes an annual covering of polythene that visitors often mistake for snowfall, but is now essential to protect the young shoots from cold and frosts. Then follows the signal for the coming of spring with the removal of the protective cover in March to reveal the dark green fields of the growing crop.

Although the potato was introduced in England in the late 1500s it was nearly 200 years before Jersey started cultivating the crop commercially. By the late 18th century it had become significant in the Island, eventually replacing the traditional fields of parsnips and the apple orchards as the island's dominant crop.

By the beginning of the 19th century, Jersey was exporting "earlies" to London; a trade that expanded very quickly. In 1809 a total of 849 tons of potatoes were sent to the markets of the English capital. By 1832, just before the formation of the Royal Jersey Agricultural & Horticultural Society, the potato export trade had increased to over 8,000 tons annually.

preparing the land…

2

Jersey Royals

- "Le Pitchage" – the team of horses were matched with the team of men who dug the ends of the rows, both for the plough to enter and leave the furrow. The horses and plough often stretched for more than thirty feet in length.

- In 1844 the introduction of guano gave a great stimulus to production. Guano is the name given to fertilizer produced from the droppings of millions of sea birds collected from the nesting sites on rocks and islands in the South Atlantic.

- Metal wheels take over from horses' hooves pulling "La Grand' Tchéthue"

Jersey Royals

The new Society encouraged farmers to concentrate on the production of early potatoes, and late season tomatoes, which were highly profitable crops in those days. On 7th December of 1833, the following resolution was recorded in the Society's minutes...

- Planting scenes around the coast of Jersey – (left) Flicquet, St Martin (below) L'Etacq, St Ouen

"That the hogshead of potatoes presented by Admiral Sir Isaac Coffin be divided into two parts; one part to be taken charge of by Colonel Le Couteur and the other by Mr Charles Bertram of Grouville and that these two gentlemen divide their respective lots into other two parts, one of each to be planted according to the plan recommended by Mr Knight and the other according to the mode practised in Jersey".

The Jersey farmer of the 1830s had much to learn about potato cultivation, but learn they did and the Island soon became one of the largest producers of early potatoes. Competition was tough and in 1838 the Society warned its members that quality must improve to match the best available in Covent Garden! By this time, island farmers were building their agricultural future on the export of early potatoes to London.

- Man-power rather than horse-power

- Une "Chuviéthe à bras"- the hand-barrow was used to carry materials over ploughed land. (farm workers at John Francis Le Brocq's farm, St Ouen, c.1935)

Jersey Royals

1768 was an important year as it saw the manufacture of a specially designed plough for Jersey – La Grand' Tchéthue – a heavy, deep-trenching plough. La Grand' Tchéthue became a community implement as it required six to eight horses to pull it, and with the small farms of those days, neighbours worked together with all their horses to do the ploughing. The days were long and hard, eased by the morning and afternoon breaks for cider and fresh vraic buns and dough cake made by their wives, and the succulent "Bean Crock" midday meal. In this way the farmers prepared their land for the potato crop.

Jersey's topography with its south-facing aspect catches the early spring sunshine making the land very suitable for growing early potatoes. Other areas of Britain, for example, Cornwall and Pembrokeshire, also grow early potatoes but Jersey's more southerly location ensures that its produce is the first on the market. However, the richness of Jersey's soil alone was not enough to feed the hungry potato. In 1844 the introduction of guano gave a great stimulus to production. Guano is the name given to fertilizer produced from the droppings of millions of sea birds collected from the nesting sites on rocks and islands in the South Atlantic.

- Planting with a tractor and winch at Gorey.

- Polythene mulch offers protection from frosts in the early growing period. (St Catherine's, c.1992)

Jersey Royals

A fungal disease known as blight first appeared in the early 1840s resulting in the near collapse of the potato industry in Jersey. In Ireland, potatoes were grown almost to the exclusion of cereals and other food crops. They formed the staple diet of the population, and when the disease devastated the potato crop, the ensuing famine caused great hardships and starvation.

It was the effects of blight on the local crop that encouraged a far-sighted and progressive St Ouen farmer, Jean Lecaudey, to realise that growing on the south facing côtils would produce a much earlier crop. Those potatoes, grown in the warm, sun-blessed slopes came to maturity in a shorter period of time and reduced the risk of blight infestation.

Lecaudey travelled widely in Britain and France to search for other early varieties, which he introduced and grew on his and neighbouring farms. The early potato industry swiftly prospered because of Lecaudey's experiments, and to show their appreciation, the farmers of Jersey formally presented him with a purse of 150 guineas, a gold watch and chain and a testimonial signed by many grateful, and by now prosperous, men.

The early potato industry was already seventy years old, when, in 1878, a major development took place; the discovery of a new potato, the Jersey Royal Fluke, with its unique flavour and texture that we know and love today.

• "Waste not, want not" Thatched cottage in St Ouen, showing potatoes growing right up to the doorstep. (c.1919)

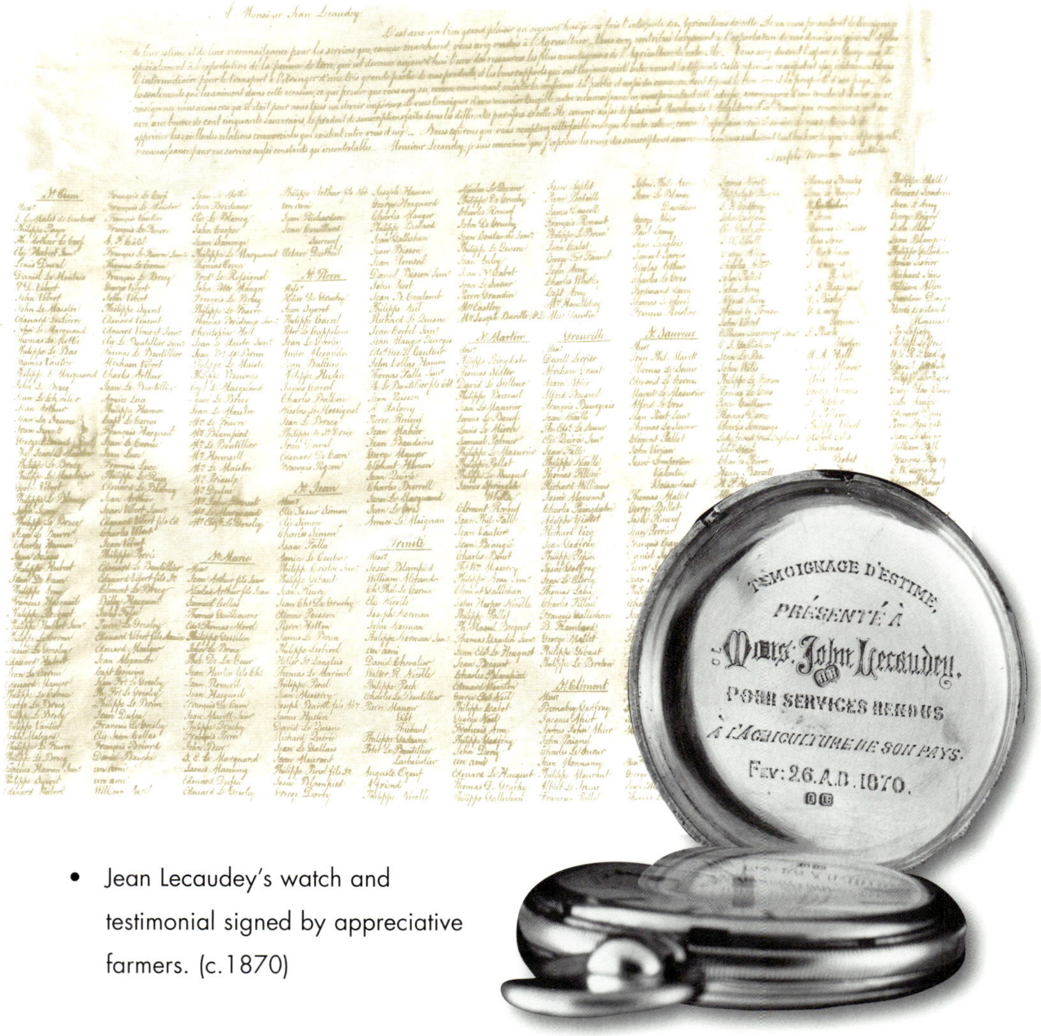

• Jean Lecaudey's watch and testimonial signed by appreciative farmers. (c.1870)

- An early scene overlooking St Brelade's Bay and Church (c.1900)

Jersey Royals

Until the "Royal" appeared, an assortment of varieties had been grown. Then, Hugh de la Haye who farmed at Bushy Farm in the parish of St Helier saw two enormous potatoes displayed on the counter of a local merchant's store on The Esplanade, owned by the variety pioneer, Jean Lecaudey. One particular potato had no less than sixteen "eyes" or "shoots" and asking for the potatoes, de la Haye took them home to conduct an experiment of his own.

This extraordinary potato was cut into sixteen pieces and planted in a côtil above Bellozane Valley.

Fifteen of the plants survived that growing season and one in particular produced kidney-shaped potatoes. Mr de la Haye carefully selected his infant crop over several seasons until there were sufficient stocks to plant commercially. The "Jersey Royal" was born!

- The Le Conte family at St Lawrence, June 1914.

- "L'Auberge du Nord", St John, when it was a farm. The Richard family in the early 1900s.

- Breton women piling the stalks and clearing the field ready for the second crop. "La fourque à chînq de"- the five-pronged fork was especially made for digging potatoes in Jersey.

Jersey Royals

Hugh de la Haye, like John Lecaudey, was later honoured by islanders at a formal gathering with a testimonial and a purse of golden sovereigns. Through the foresight and endeavours of these outstanding men the prosperity of the Island was guaranteed for the next hundred years.

By the beginning of the 20th century, the pattern of farming had settled into a cordial partnership between the Jersey Royal and the Jersey cow. They complimented each other well as the lifting of the "earlies" allowed time in the later part of the year for a crop of roots to be grown. This fed the cow during the winter when she returned the compliment by supplying much needed fertilizer for the hungry soil.

The area of land growing potatoes grew rapidly until over 20,000 vergées per year was devoted to the crop; about one-third of all land in the Island.

The potato was now a vital part of the Island economy; yesterday's smallholders had become farmers who derived a good living from the land. They now employed people to harvest the potatoes, even on the smallest farms, and they prospered.

The workforce at that time came from France, mostly from Brittany and a few from Normandy. Some 4,000 Breton men and women arrived each year by boat from Lessay, Carteret and La Hague and also from villages in the Cote du Nord areas such as Ploeuc, Binic and St Brieuc. They stayed for "the season" and worked in teams of three, called "le tâcheron" or "fork". One man dug the potatoes with the fork, and a second, who was the "shaker";

• Breton workers arriving for the Season

• "Le Tâcheron avec sa fourque compliéter" – the team that consisted of the complete 'fork'.

• "La bouochie" – morning's welcome break.

• Saturday afternoon after a hard week's work

• Rozel showing the south-facing côtils (c.1919)

Jersey Royals

shook the potato stalk of its fruit, and the third followed behind with two baskets to grade and pick up the valuable crop.

With growing unemployment in the United Kingdom during the Great Depression of the 1930s, the English government requested that the Island banned the entry of Breton labourers and employ English ones in their place. During the following years, until World War II,

the labour force on the farms spoke the English language, albeit with a variety of accents including the soft Irish lilt, rather than French.

Little did Jerseymen know, in 1939, what lay ahead in the following years and the farmers continued to plant their crop of early potatoes in the prepared ground. A total of 22,099 vergées was planted, an area never before grown, and never again exceeded. That year, Jersey headed the list of potato imports into the United Kingdom with 67,738 tons, followed by Spain with 14,700 tons and the Canary Islands with 6,900 tons.

- The season in full swing using four-cabot barrels. The Ozouf family at Le Tapon, St Saviour in May 1949.

Jersey Royals

The five years of the German Occupation of the Channel Islands saw an invasion of another sort; the Colorado potato beetle. It established itself in several fields, having travelled on the incoming tides from France. While the little black and yellow-striped beetle caused alarm in the community, fortunately the campaigns in the years immediately following the occupation were very successful in eliminating this particular enemy.

• John de la Haye and his farm workers at West Lynn, St Ouen. (c.1930)

The ending of the war saw resumption in the potato trade but on a reduced scale. In 1946 Breton labourers were allowed to return to Jersey but with more work becoming available in their homeland, the numbers travelling were less than in the heady days of the 1920s.

Changes in farming were obvious everywhere. By 1950 there were more tractors than horses; and lorries had largely replaced the horses and vans. Loads of potatoes moving from the outlying country parishes to the harbour in St Helier were running on rubber tyres instead of iron. A potato harvester, a common sight in the enlarged fields nowadays, first appeared in the Island in a much simpler form in 1939. The first decade after the war saw prosperity return to the countryside, but the decline in numbers

• Lunch-time and time for a picture!

of farms was inevitable. In 1950 there were over 2,000 farm holdings; fifty years later there were less than a quarter of that number. But even today, more than 19,000 vergées of early potatoes are grown each year.

Fortunes have changed; today the language of the workers in the fields is not French, but Portuguese, perhaps even Polish, while most of the hard work of potato digging is done standing on the harvester rather than kneeling over the weighty baskets of yesterday. The U.K. demand for the early potato is still strong with from 35,000 to 45,000 tons, depending on the season, being exported annually, but the competition from further afield increases with each new season.

Today's farms are larger, with sophisticated computerised machinery to assist with the growing, harvesting and exporting of the potatoes. The cycle of the Jersey Royal remains constant and each year, from April to June, the countryside comes alive as the "Royals" are lifted.

- Forty years of progress.
(left to right) The Le Quesne family, La Retraite, St Saviour in the 1960's;
digging with a typical harvester at St Ouen in the 1990s;
a modern harvester at Vinchelez, St Ouen.

Jersey Royals

Jersey is situated a little over one hundred miles south of England and supplying the early potatoes onto the British market in pristine condition has always been a challenge to the industry.

In the very early years sailing ships carried the crop across the channel and it was not until the mid-19th century that they were replaced by steamships that took the potatoes to England's south coast ports for onward journey by freight trains to the inland markets.

The journey started when the farmer took his "load", firstly by horse and van, then by lorry, to the Weighbridge at St Helier's harbour. From early times, the Weighbridge has always been a hive of activity, especially when the selling of the potatoes by the farmer to the merchant became a talent in itself.

The fortunes of selling potatoes "on the bridge" were very mixed and differed each season. The loads would cross the weighbridge where officials weighed the vehicle and its contents. As it left the scales, buyers representing some of the thirty-or-so merchants surrounded the farmer and his load. Prices offered would vary depending on the quality of the potatoes, and on the market at that time. Before the 1930s when the telephone link was established with the UK, communications

• Packing the potatoes

- Shipping in the 19th century.

Jersey Royals

were all done by telegram. The merchants were advised of the day's opening and expected prices by the mainland buyers and at the end of each day they confirmed their purchases. When the load was sold and unloaded in the merchant's store, the empty vehicle was weighed and the farmer given a receipt with the net weight of his produce. These receipts became currency and could be deposited in a bank for collection or turned over to one of the merchants who supplied the farmer with his seed and fertilizer, as payment.

For a variety of reasons, prices fluctuated from hour to hour, and sometimes even from minute to minute. Farmers also had the choice of taking the average price of the day, which was less of a gamble.

- "The waiting game…"

- The biggest load of potatoes ever to cross the Weighbridge! – Mick and Edward A'Court with son Ron at L'Etacq.

- The busy Esplanade. Note Policemen's white summer helmets.

- "Over the Bridge – second time round for nett results". To establish the weight of the potatoes the full load, including horse and man were weighed, and then weighed again empty.

Jersey Royals

• Then and Now – 1939 – 2001

Jersey Royals

• Seed-boxes

Today, marketing groups supply the supermarket chains and wholesalers on the mainland and handle the marketing of the Jersey Royal potato. The last job of the farmer's potato season is to harvest the seed for the following year, and he pays particular attention to this task. He is meticulous in the selection of his seed potatoes, and usually identifies particular fields that are known for producing the best stock.

The seed is stored in potato boxes, or chitting trays, for the rest of the summer under carefully controlled temperature, humidity and lighting conditions. In November, the season really begins with "standing" the potatoes. This is part of a carefully controlled process when the farmer sorts and discards any rotten or poor potatoes and "stands" the best ones in boxes according to size.

"Standing," means the removal of the first single shoot by rubbing it off the potato, which is then placed upright, "eyes up" in the chitting trays. This controls the development of the potato, and increases the quantity of early potatoes produced from each plant. The boxes of "stood" potatoes are then stacked in the potato store until planting time. This is one part of the potato story that has changed very little and is still done by hand.

• Standing potatoes has changed little over the centuries.

• In the winter months, farm sheds are stocked high with "Stood" potatoes

• All the family celebrate as the last of "La s'menche", or seed, is lifted at Highstead, St Saviour, 1934.

• Archirondel (c.19th century)

Jersey Royals

The earliest of the crop is to be found on the steeply sloping south-facing côtils in the coastal areas of L'Etacq and Rozel in particular and they provide a wonderful backcloth to stunning scenery. Today, these côtils are worked mostly by tractor and plough, using a winch, and the crop is harvested with a light digging plough. Some of the smaller patches of côtil are still dug by hand with the potatoes being lifted using the fork in the time-honoured way with "le tâcheron".

The preparation of the côtils starts in December when they are ploughed using the tractor and small winch. The tractor parked at the top of the field hauls the plough, guided by a labourer, up the steep slopes. Reaching the top, the man on the plough runs down the côtil with it and starts over again. It is a time-consuming process but with efficient and energetic labour more than two vergées a day can be prepared in this way.
The seed potatoes are planted in early January and covered with sheets of polythene mulch. This ensures a "greenhouse" type environment keeping the delicate potatoes warm and safe from all but the severest frosts and allows them to "breathe" through the perforated plastic.

Towards the end of February or early March, depending on the season, the plastic is removed, allowing the crop to mature in the early spring weather. By the first week of April, the farmer starts to lift the high-value crop.

- "The first of the crop" – digging at Portelet in fields farmed by Wilfred de Gruchy.

- The côtils are still dug by hand to protect the soft skin of the first earlies. Scenes at the Vautier family farm at Rozel.

Jersey Royals

Jersey has some of the most formidable tidal flows in the world, and the strong movement of the sea deposits large quantities of vraic, the local word for seaweed, on its shores. Vraic has been used on the land in Jersey since the twelfth century, or earlier, and before the importation of guano and other fertilizers in the mid 19th century, the farmers only used farm manure and vraic to improve the soil and fertilize the crops.

• Every little counts!

In the past, vraicing used to be a full-time occupation for some families, and stacks of seaweed would be dotted on Les Mielles, the sand-dunes along St Ouen's bay in the west of the island, ready for sale to needy farmers. Most farmers, however, chose to collect their own and would spend many days throughout the year travelling between their fields and the beach to harvest the valuable vraic.

Laws governed the collection of vraic; for instance, no vraicing was allowed on Sundays, between Saturday sunset and Monday sunrise. Farmers had to wait for the tide to recede to a certain point, indicated by poles on rocks at La Pulente, before they could proceed to the water's edge. It was a common sight for dozens of horses and carts to be lined up at the top of St Ouen's bay, waiting to speed down to favourite spots as soon as the sea reached the low watermark. Horses used for vraicing were not generally the best farm horses, but a good, fast

• La Rocque, St Clement

- Stacks of vraic at St Ouen's Bay.

- Overlooking St Ouen's Bay from the Corbiere.

• Le Pulec, L'Etacq

• Piled High,
Le Hocq

Jersey Royals

one could just give the edge to his owner by being the among the first to reach the best spot. The gatherers had to mark their area alongside each other; a "channel" between six and ten feet wide, heading straight out to sea. Making piles of seaweed created these marks, high enough for the receding tide not to topple them, and woe betide anyone not making their marker line straight!

According to the Code of Laws published in 1771 (when all laws of Jersey were, for the first time, actually printed), all working land-owning men resident in the six western parishes of St Ouen, St Peter, St Brelade, St Lawrence, St John and St Mary, were allowed a certain number of "lots" of vraic, based on the area of land they farmed. For instance, a man with between three and eight vergées could have one and a half lots, whereas a man with over 60 vergées could have six lots. Dates for the cutting of vraic with small sickles, as opposed to the collection of drift vraic, were strictly controlled. On no account was vraic to be torn from the rocks, as this harmed the plants and impaired later growth.

• Harvesting the sea – vraicing on a spring tide during the 1930s.

Jersey Royals

In the east of the island, where the coastline is quite different and the sea is more gentle than the exposed westerly waves of St Ouen, the laws of collecting and marking were different. In the 17th century, the Payn family who were the Seigneurs de la Malletière, were granted the privilege of "une pêcherie à vraic et à poisson", fishing and vraicing rights in an area near a rock called L'Avarison in the Bay of Grouville. These rights had long been enjoyed by the Seigneurs but were established by a judgement of the Royal Court of 28th April 1747, for which the owner had to pay a "rente" to the sovereign. The limits of the rights were specified by the court and today one can still see the letter "P" clearly marked on six rocks covering a circular area of about one and a quarter miles.

There are more than 240 types of know algae around the island's shores, but this variety was of little concern to the farmers. Certain species identified by colour were used for certain crops. Green vraic, known as "cliaque" was favoured for spreading on grassland in the spring. This variety was cut from the rocks following the spring tides, and spread on the grass. The majority of vraic gathered and used on potato and other crop fields was know as "collet" by the farmers from the west of the island, while the easterners called it "collain". This was the brown vraic that the sea delivered in copious quantities daily. The horses and carts were backed into the sea so that farmers could haul in the seaweed with large pronged rakes, and then fork it into their carts.
Some farmers mixed the vraic with stable manure and hedge cuttings in the farm manure heap in a form of

- Techniques may change...

- Spreading on the land , Le Clos Marquand, Mont Matthieu in 1973.

• Low tide at La Rocco.

• Vraic for sale at La Rocque

compost, but more often vraic was spread directly onto fields to "breeze", a form of shallow ploughing, then deep-ploughing it into the soil. Before the 19th century, many poor families along the coast collected the drift vraic and burned the seaweed to produce vraic-ash that was particularly valuable for the cultivation of wheat.

Later in the 19th century and early part of the 20th century, vraic gathering would involve all the family, who would go down to the shore, the men dragging and loading the heavy wet vraic on the carts while the women and children would collect the black winkles hidden under the weed. They would also go sand-eeling, fishing for the tasty little fish near the water's edge. The wives prepared a huge supper of vraic-buns, dough-cake and the like to be washed down with cider and everyone enjoyed the occasion. As one retired farmer commented, " it was like a town at night down at La Rocque!" Those were the days!

In the 1930s, lorries gradually took over from the horse and cart, although during the Occupation, farmers had to revert to "equine engines" as they carefully negotiated their way through the land mines in St Ouen's bay to retrieve the sea's harvest. Following the War, the "boom years" of vraicing returned, but the practice fell from favour when manufactured fertilizers became more easily available. The tradition is still carried on, albeit in a much reduced way today, with some farmers preferring to use Jersey's natural fertilizer and soil conditioner from the sea for their early potato crop.

• St Catherine's Bay

- Each their own "pèrtchi", or patch.

- "Au lanchon" – sand-eeling.

Jersey Royals

part 3

Tomatoes and Flowers

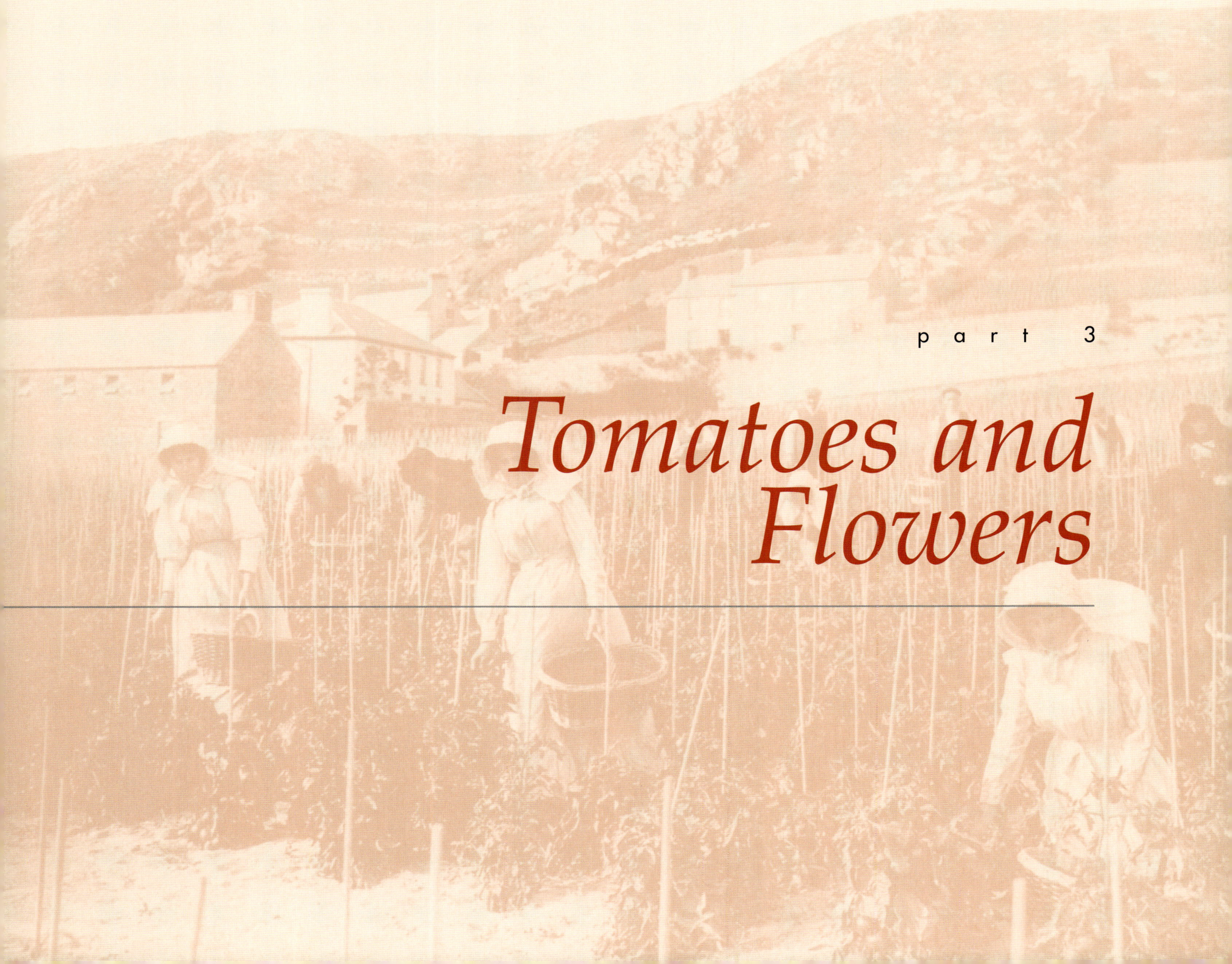

Tomatoes and Flowers

The gently sloping vineyards of France and Italy look splendid with the rows of growing vines, but Jersey, some years ago had a similar appearance, not with the grape, but fields of outdoor tomatoes.

The history of the Island's tomato crop may not be as long, or as significant, as the Jersey Royal potato, but it still had its place in the fortunes of many Jersey farmers. Although tomatoes had been raised under glass for some time, it was not until the late 19th century that farmers started growing tomatoes in the light, sandy coastal land of Jersey.

Many varieties have been favoured by different growers, including the Devon, the Flying Dutch, the Jambart Beauty and the Potato Leaf, but the one that was probably the most popular was the Sunrise, with its thin skin and tasty juicy centre.

In the early 1900s, a little over two thousand tons of tomatoes were exported each year to the mainland, but by the outbreak of World War II, a total of over 30,000 tons, grown on nearly 7,000 vergées of land, was shipped annually.

The versatility of the tomato attracted farmers, as they could be grown either as a first or second crop following the lifting of the early potatoes on a wide variety of soil types. This ensured extended work for the Breton labourers on many farms. Instead of working for the relatively short potato season, there was work to keep them in the Island

• The 'vineyards of Jersey', the outdoor tomato fields.

• Field at Pied du Côtil, St Andrew's Road, First Tower.

Tomatoes and Flowers

until the last of the fruit was picked in the autumn. In between work on the two field crops, the hay and corn was harvested. Root crops were sown and weeded, for the winter cattle feed and the branchage, the trimming of Jersey's hedgerows, all cut with razor-sharp sickles, gave the island its bi-annual tidy-up in July and September.

Tomato cultivation was very labour-intensive. From the sowing of the seeds to the picking of the fruit, constant care and attention was required. The young plants were raised in trays, under glass, before being "pricked out". In time the best plants were eventually planted out in rows and staked with bamboos, or timber "crosses" to support strings or wires to which the plants were tied. Throughout the growing period, the ground was weeded regularly by horse and hand hoes, and the plants sprayed regularly against blight.

As the tomato plants grew, the main shoots were secured with

• Spraying tomatoes with copper and bluestone against fungal blight

• A fine truss of tomatoes! (c. 1938)

- Pricking out the seedling tomatoes, Le Tapon, St Saviour, May 1949.

Tomatoes and Flowers

string to the support. To encourage a better quality crop, the side shoots were pinched out, and to stop the plant growing too tall, the growing tip was removed after the third truss set. This gave strong trusses of quality tomatoes and ensured even ripening.

Regular picking was a top priority, so that the picked tomatoes were in the best condition for the market. The picked fruit was then sorted, graded and packed in trays for transport to St Helier harbour for loading and eventual shipment to England.

Following the German Occupation, which stopped the export of any of Jersey's crops, farmers returned to growing Jersey "Royals" and "Toms". In 1948, a total of 35,000 tons of tomatoes were exported, with a value of £3 million, nearly three times the value of the early potato crop at that time. Those were gloriously "red days" for the Island farmer!

Sadly, it did not last. The increasing cost of labour, competition from foreign produce, and the growing influence of the supermarkets setting their sights on cheaper imports, have significantly reduced the outdoor tomato crop of Jersey.

Today, some of Jersey's light sandy soil near the coast is still used to grow the vines of "love apples" with that succulent fruit providing a distinctive flavour on the summer salad table.

• Planting tomatoes.

- Grading and packing tomatoes in 'dutch' trays at Les Tours, St Clement.

- The whole family helps with the picking.

• Grading tomatoes at Les Tours, St Clement.

Picking and packing in the field – the Simon family at Val de la Mare, St Peter, 1953.

• Tomato picking at Les Niemes Farm, St Peter. The Morin family who worked for Mr Le Rossignol. (c.1925)

- Each year at the Lavender Farm at St Brelade, a blaze of colour attracts the eye. The Christie family have built a good business with their flowers on a farm that was once the home of the noted "Natalie" cattle family.

- The Island has a good flower export industry and a large variety of cut flowers as well as bulbs have traditionally been sent to the United Kingdom and the European continent for many years. More recently, a local company, Flying Flowers has become the world's largest exporters of cut flowers by mail – going to every part of the world.

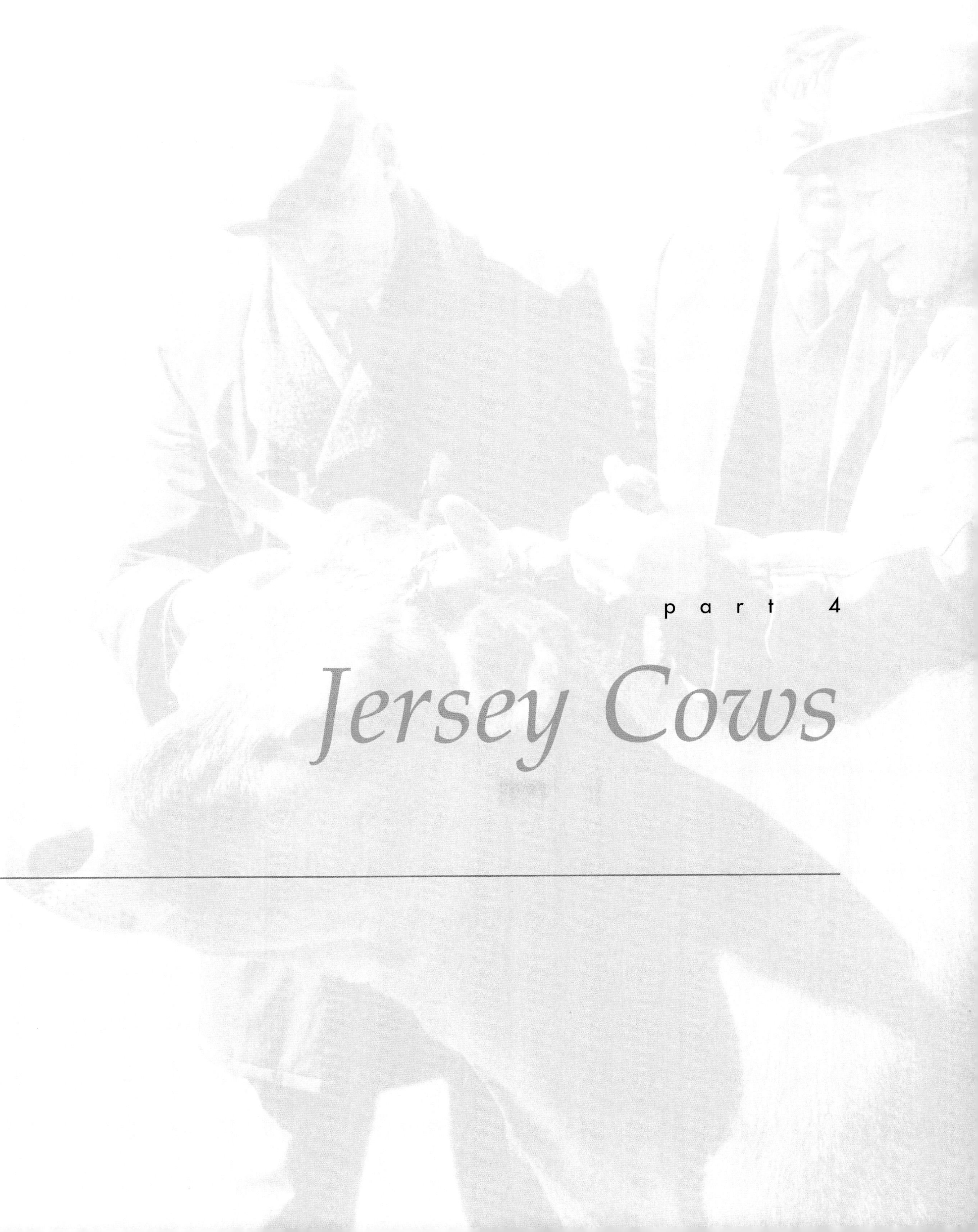

part 4

Jersey Cows

Jersey Cows

"The treasure highest in a Jerseyman's estimation is his cow. She seems to be a constant object of his thoughts and attention; and that she certainly deserves."

This was the view of Thomas Quayle in 1812 when he visited Jersey on behalf of the English government to study the general state of property, cultivation and rural and political economy of the island.

Pride of the Island farmer and without doubt, its most famous envoy; the Jersey cow has travelled from her native shores to new homes across the world.

Today, Jersey cattle graze in all continents, and most individual countries, all of which have a direct hereditary link to the Island. Numerically, the Jersey is the second largest breed of dairy cattle in the world.

As the Jersey cow developed in its Island home, the growing popularity of the breed demanded legislation to protect its purity. The first Act of Legislation in 1763, prohibited cattle from France coming to Jersey, then being re-exported to England as "Alderneys". This pseudo breed was named after the northernmost Channel Island. Pacquet ships carrying provisions and mail for all the islands left Southampton, travelled to Jersey where they would take some cows on board, then to Guernsey, and lastly to Alderney before returning to England. The ship's captain would then record the number of cattle to be landed, and they were all referred to as Alderneys, although it was most

- Hamptonne, St Lawrence

- One man and his cow... Brampton Farm, St Ouen (c.1932)

- Roseland herd of Carlyle Le Gallais shortly after the Occupation.

- "The treasure highest in a Jerseyman's estimation….."

unlikely that very many or even any of them would have called that small island their original home.

The States of Jersey amended the law several times, and finally in 1878, all cattle imports, from any source, were prohibited. By this time, the Jersey cow was renowned all over the world for its high quality milk providing the best butter and other products.

The 19th century proved a prolific period in the development of the Jersey cow internationally. Such was the demand for Jerseys in the early 1800s that in 1833 a group of far-seeing, progressive agriculturalists formed the Royal Jersey Agricultural and Horticultural Society, and a year later, the first scale of points linked to parts of the animal was introduced to make improvements to the breed.

Two of the best cows were selected as models, one was considered to be perfect in her forequarters and barrel, and the other in her hindquarters. With the help of the best breeders and dealers, a scale of points for the guidance of judges at the shows was drawn up. This scale of points, modified occasionally, is still the benchmark for judging cattle to this day.

• Country contentment! A peaceful summer's day.

- Nelson Renouf's herd at Bel Val, St Martin, overlooking St Catherine's.

- Home to be milked … two generations of John Germain in the 1950s.

• Broadfields, St Lawrence in the 19th century.

Jersey Cows

• "Early Herd book Judges" – this was part of the RJA&HS Centenary Pageant in 1933, reflecting earlier times.

The fundamental principle of the breeding of pedigree dairy cattle is to have authenticated records and in Jersey, the Jersey Herd Book is responsible for this function.

The founding of the Jersey Herd Book in 1866, set out to register the birth of each heifer and bull calf, with a qualification of that animal following her first calving, in the case of heifers, and at a year old for bulls, when they were ready for service.

The "Herd Book Exams" were a unique feature of the Island's cattle breeding industry and Jersey was the only place in the world where this tradition was continued, for over 100 years, until 1972 when classification replaced qualification. In those days, before artificial insemination and the international exchange of genetics, the Herd Book examinations enabled the Island Jersey to make faster genetic gains than other countries.

Six times each year, a group of breeders, including the Herd Book judges, office staff and officials of the Royal Jersey Agricultural Society travelled around the island, stopping at designated places where farmers brought their cattle to be inspected and qualified for acceptance in the Herd Book. In the early years, the group travelled by horse and carriage,

• The first entries of the Jersey Herd Book, published in the local newspaper. The Herd Book forms the official registry of all Jersey cattle born in the Island and is maintained to this day. Registered Jersey cattle in other countries can trace their pedigrees back to the Jersey Herd Book. The first bull and cow to be registered in the Jersey Herd Book were "Dandy", owned by Mr James Godfrey of St Martin, and "Daisy" owned by Mr P Paisnel of St Clement.

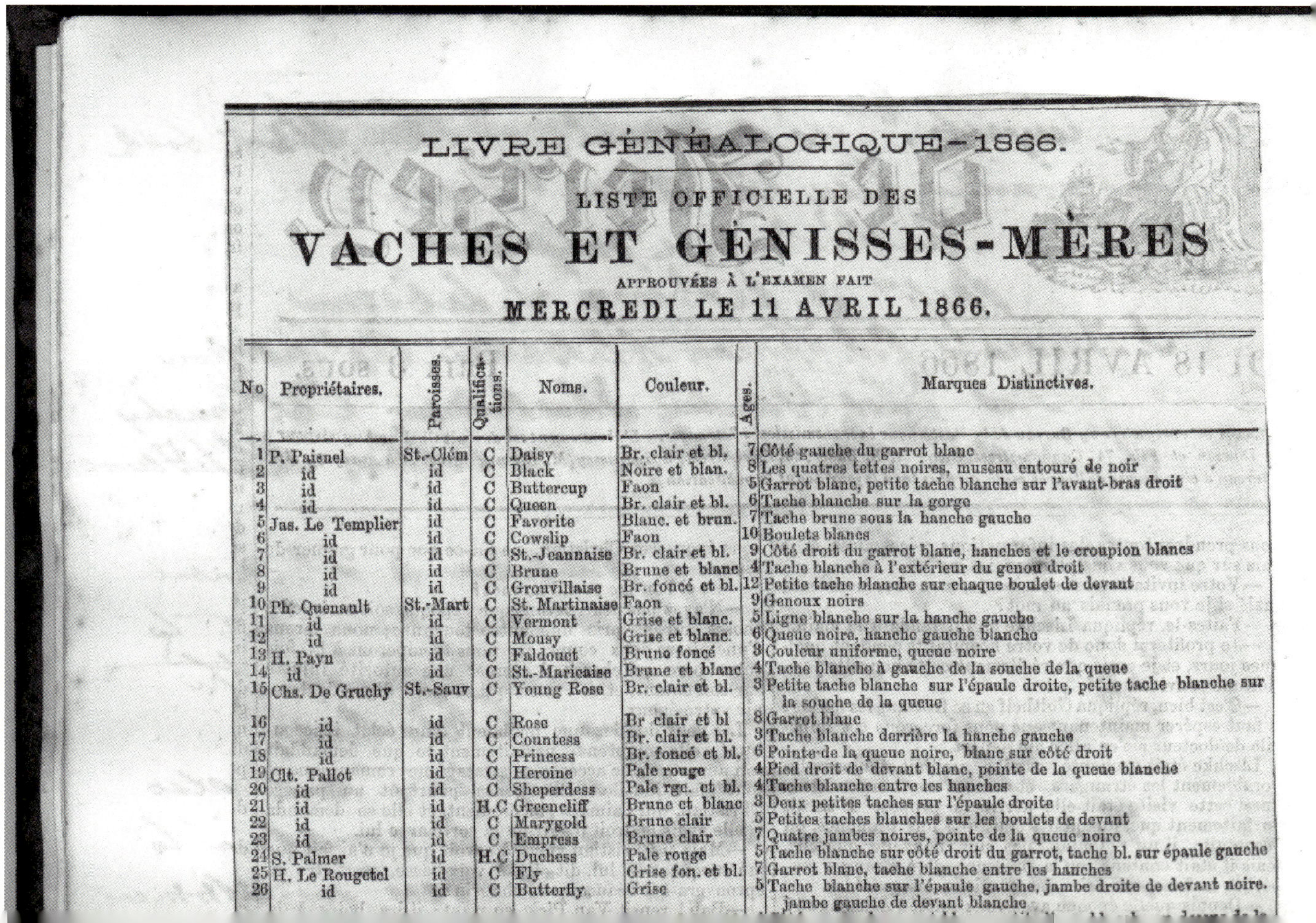

LIVRE GÉNÉALOGIQUE—1866.

LISTE OFFICIELLE DES

VACHES ET GÉNISSES-MÈRES

APPROUVÉES À L'EXAMEN FAIT

MERCREDI LE 11 AVRIL 1866.

No	Propriétaires.	Paroisses.	Qualifications.	Noms.	Couleur.	Ages.	Marques Distinctives.
1	P. Paisnel	St.-Clém	C	Daisy	Br. clair et bl.	7	Côté gauche du garrot blanc
2	id	id	C	Black	Noire et blan.	8	Les quatres tettes noires, museau entouré de noir
3	id	id	C	Buttercup	Faon	5	Garrot blanc, petite tache blanche sur l'avant-bras droit
4	id	id	C	Queen	Br. clair et bl.	6	Tache blanche sur la gorge
5	Jas. Le Templier	id	C	Favorite	Blanc. et brun.	7	Tache brune sous la hanche gauche
6	id	id	C	Cowslip	Faon	10	Boulets blancs
7	id	id	C	St.-Jeannaise	Br. clair et bl.	9	Côté droit du garrot blanc, hanches et le croupion blancs
8	id	id	C	Brune	Brune et blanc	4	Tache blanche à l'extérieur du genou droit
9	id	id	C	Grouvillaise	Br. foncé et bl.	12	Petite tache blanche sur chaque boulet de devant
10	Ph. Quenault	St.-Mart	C	St. Martinaise	Faon	9	Genoux noirs
11	id	id	C	Vermont	Grise et blanc.	5	Ligne blanche sur la hanche gauche
12	id	id	C	Mousy	Grise et blanc.	6	Queue noire, hanche gauche blanche
13	H. Payn	id	C	Faldouet	Brune foncé	3	Couleur uniforme, queue noire
14	id	id	C	St.-Maricaise	Brune et blanc	4	Tache blanche à gauche de la souche de la queue
15	Chs. De Gruchy	St.-Sauv	C	Young Rose	Br. clair et bl.	3	Petite tache blanche sur l'épaule droite, petite tache blanche sur la souche de la queue
16	id	id	C	Rose	Br clair et bl	8	Garrot blanc
17	id	id	C	Countess	Br. clair et bl.	3	Tache blanche derrière la hanche gauche
18	id	id	C	Princess	Br. foncé et bl.	6	Pointe de la queue noire, blanc sur côté droit
19	Clt. Pallot	id	C	Heroine	Pale rouge	4	Pied droit de devant blanc, pointe de la queue blanche
20	id	id	C	Sheperdess	Pale rge. et bl.	4	Tache blanche entre les hanches
21	id	id	H.C	Greencliff	Brune et blanc	3	Deux petites taches sur l'épaule droite
22	id	id	C	Marygold	Brune clair	5	Petites taches blanches sur les boulets de devant
23	id	id	C	Empress	Brune clair	7	Quatre jambes noires, pointe de la queue noire
24	S. Palmer	id	H.C	Duchess	Pale rouge	5	Tache blanche sur côté droit du garrot, tache bl. sur épaule gauche
25	H. Le Rougetel	id	C	Fly	Grise fon. et bl.	7	Garrot blanc, tache blanche entre les hanches
26	id	id	C	Butterfly	Grise	5	Tache blanche sur l'épaule gauche, jambe droite de devant noire. jambe gauche de devant blanche

- The first scale of points for a Jersey bull approved by the RJA&HS on 5th January 1850. This was drawn up by Colonel Sir John le Couteur, whose portrait appears opposite part 6 of this book. An early pioneer of the Society, he was its first Secretary and later President on four occasions.

- Herd Book exam in progress at Carrefour Selous, St Lawrence (c.1954)

replaced later by cars that always included a Curwoods taxi to be used as the mobile office.

These were interesting times and the Herd Book party attracted quite a number of followers, including local cattle exporters who would accompany the group to keep a watchful eye for any new young "bovine starlet" that could be purchased for later sale overseas. The group was well hosted too! Once during the morning and again in the afternoon, the group were treated to refreshments by ladies at one of the farms; tea, or coffee and home-baked cakes, and always the offer of a "little tot" to keep one warm! Lunch was taken at St John's Hotel, in the early days and later, La Belle Etoile in St John, the halfway point on the Island tour.

The importance of these Herd Book exams cannot be underestimated, as it was the hub upon which breed improvement was made. When the judges appraised an animal the farmer was given, either a white card, for Highly Commended, or a red card, for Commended, the lesser of the two distinctions. The Secretary of the Jersey Herd Book was then authorized to assign a Herd Book number, with the grade designation, to the animal so it could officially become a parent of the next generation. This judgement set the standard by which the breed improved.

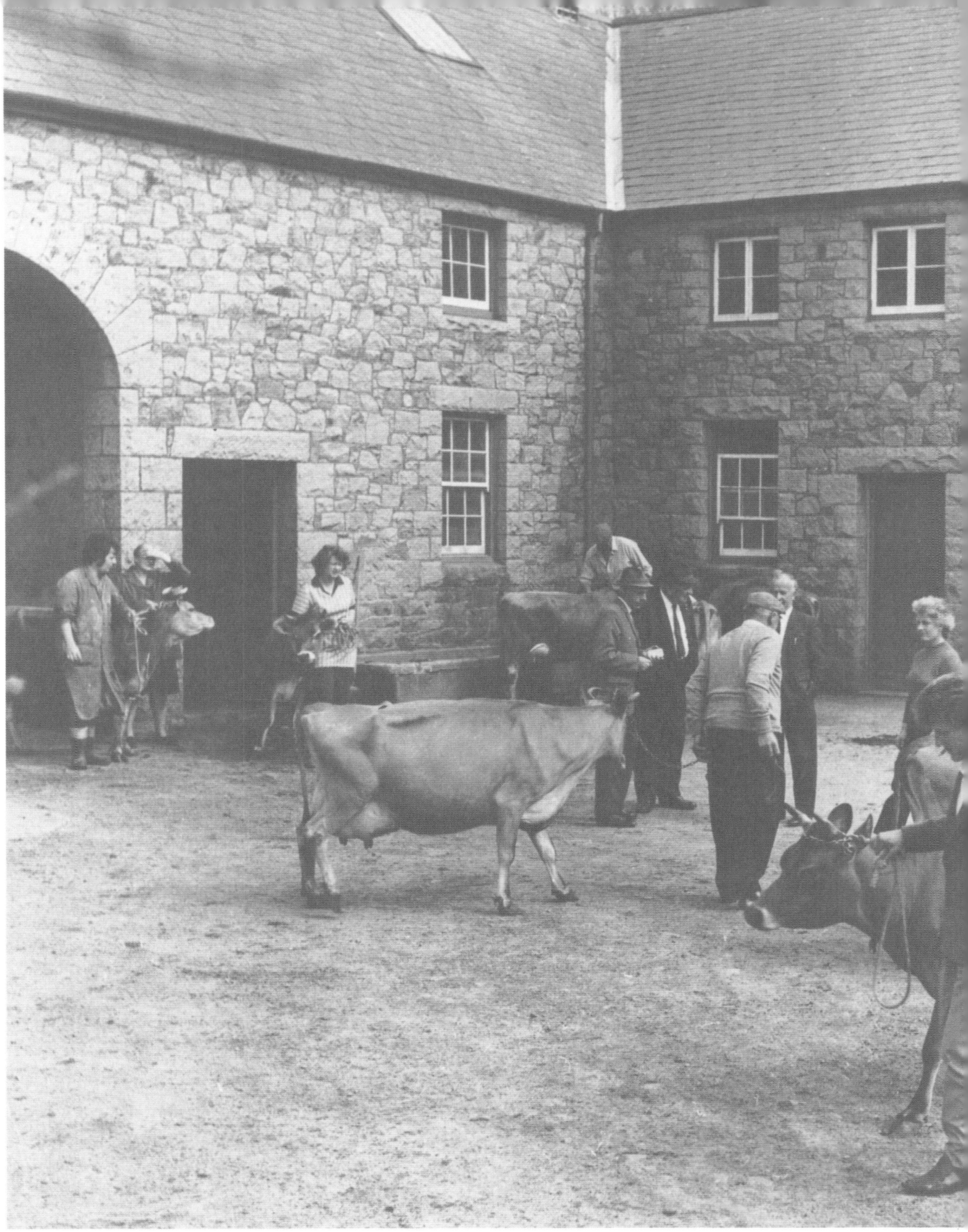

• Heads I win!

• Herd Book exam in progress at Manor Farm, St Peter (c.1970)

• Herd Book judges inspect the dam of a young bull. (c.1970)

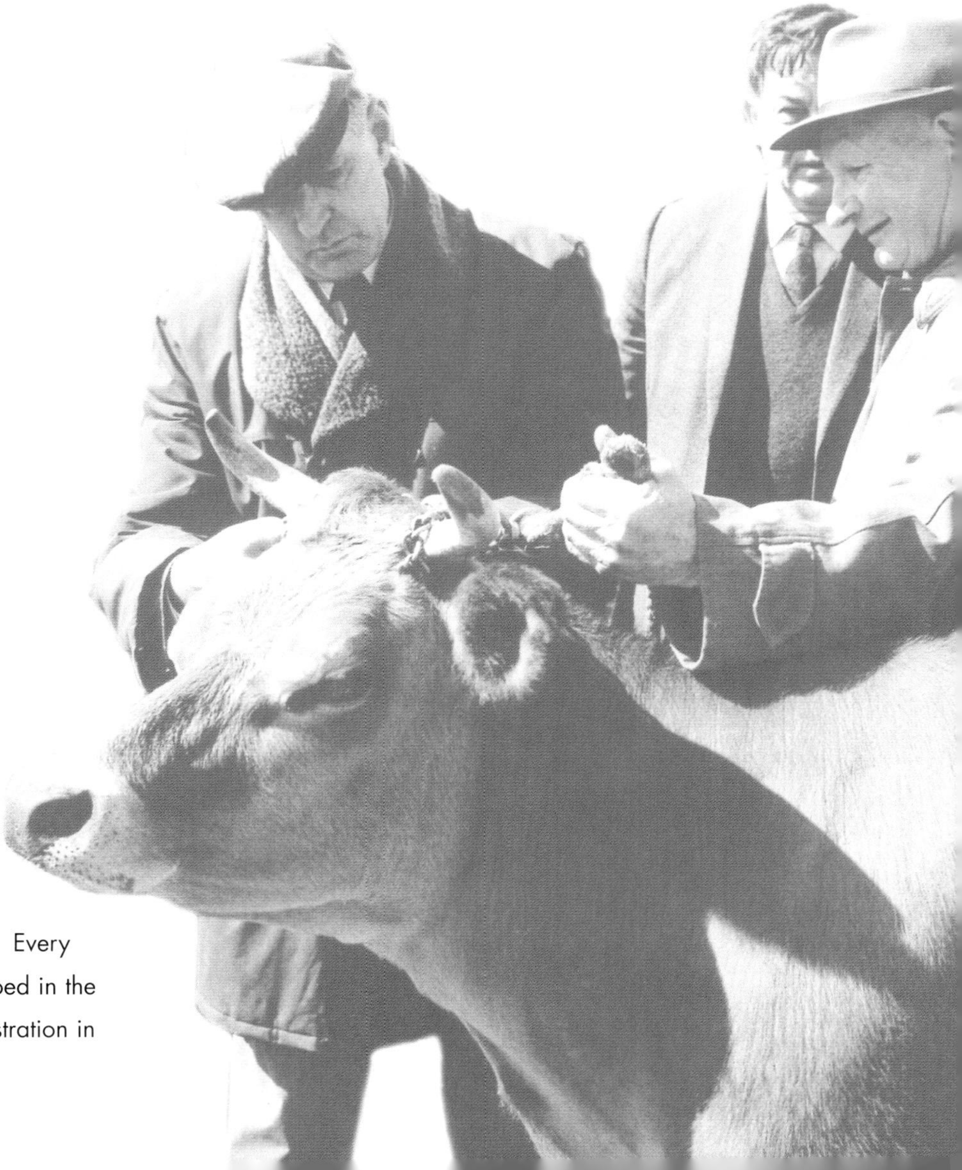

• Checking the tattoo number. Every 'Jersey' in the Island is tattooed in the right ear, identifying its registration in the Jersey Herd Book.

Jersey Cows

Jersey cattle have been bred in the Island for centuries, but it was in 1866 when the Jersey Herd Book was formed, that improvement of the breed could begin in earnest. Through the registration and comparison of individual animals, selection of the best bloodlines could be made. This process has been repeated across many species and breeds of livestock.

The breeding of cattle by blending bloodlines is an art at which many Jersey farming families became noted for their expertise. This knowledge has been passed down the generations and has made the breed the success it is today.

The pictures on this page show the intertwined story of farming families and cow families.

The man in the centre is J.P.Marett, one of the very early breeders of Jersey cattle. He was famed for breeding two particular animals; the first was the cow Sultane, registered as No. 7 in the Jersey Herd Book. She can be described as the "mother" of the Jersey breed as virtually every Jersey in the world will trace at least one line of its pedigree to that cow. The other is a bull named Golden Lad who was related on both sides of his pedigree to Sultane. He became one the breed's finest sires and again appears in virtually every Jersey animal's pedigree.

The gentleman in the bottom left corner is John A. Perrée of Oaklands, St Saviour, who was a famous breeder and leader of the industry, standing with a bronze sculpture of a bull known as Golden Fern's Noble. This was a gift from a very satisfied American cattle breeder who had purchased the bull and been very successful with his progeny. The picture of Golden Fern's Noble is bottom right, and he is a son of the dark coloured bull above, Noble of Oaklands, bred by Mr Perrée, one of the most successful breeding bulls of his day who was known as "King of the Island".

The cow at the top of the page is Lady Viola, the mother of Noble of Oaklands. This cow was Champion over Jersey in 1905. She was sold to England where she was unbeaten in the show ring wherever she was exhibited for the following three years. The immense respect for this cow, which was described at the time as "the best cow ever seen", prompted her English owner to export her to America where she was auctioned for the record price for a cow of $7,000. At the same sale, her son "the King of the Island" sold for a world-record price for a bull of $15,000!

Lady Viola was indeed the finest cow in the early story of Jersey cattle breeding, and her pedigree is rich in the bloodlines of animals bred by J.P. Marett forty years earlier. These gentlemen and their animals perhaps provide a flavour of the important history of the Jersey breed and the people that developed it.

Schreiber

Jersey Cows

Methods of assessing the amount of milk produced by a cow have been in existence for more than 120 years. Today, production recording is sophisticated, computers providing a range of information to assist the farmer to improve his livestock, but in Jersey in the 1880s, such advantages were not available. In 1889 the first tests to prove the dairy qualities of Jersey cows were held, to show the highest quantity of cream from a given number of milkings. Only four animals competed, the yield of three milkings being taken; the cream of the first two was subjected to analysis and that of the third made directly into butter.

Just four years later, in 1893, 24-hour "Butter Tests" which were to prove so beneficial to the Jersey breed, were held at the Cattle Market in Minden Place where the car park is today. Farmers brought their cows to the market where they were stalled, to be milked by milk maidens employed for the task. The cows producing the greatest amount of butter were awarded the gold medal, followed by the silver and bronze medal winners.

In 1912, Mr John A Perree, the Secretary of the Royal Jersey Agricultural & Horticultural Society and a prolific exporter of cattle, noted that foreign buyers were increasingly demanding authentic information about the production of individual cows, and urged the Society to introduce official

- Charles Mourant of La Croix ès Mottes, St Saviour, breeder of the famous Sybil's Gamboge, with some of the medals won by his noted herd.

- A typical dairy cow of the 1880s. This cow, Violette, was a show winner and produced over three pounds of butter in the 24 hour Butter Tests.

A "MASTERPIECE" in the Art of Jersey Breeding

Sybil's Gamboge

Bred by Chas. Mourant, St. Saviour's, Island of Jersey.

Imported and sold at Butler's 1919 auction at the world's record price of $65,000.
Winner of first prize and Peer Perpetual Challenge Cup, with his progeny, at Royal Show, in 1919.
Third prize with his progeny at Royal Show, 1918.
Twenty of his "get" sold at auction in 1919 for an average of $3,245; and twenty of his "get"
sold in 1920 for an average of $4,210; an unequalled record.
Sire of 40 Official Island test cows and 20 in Register of Merit:
Sybil's Majesty Lorraine, Register of Merit, 835 lbs. butter in 343 days.
Sybil's Bayleaf, Register of Merit, 825 lbs. butter, 11,950 lbs. milk.
Sybil's Bay Girl, Register of Merit, 748 lbs. butter, 9676 lbs. milk.
Gamboge's Combina, Register of Merit test, 716.06 lbs. butter from 12,355.5 lbs. milk, 365 days.
Sea Lass' Sybil, Register of Merit, 664 lbs. butter, 10,481 lbs. milk.
Sybil's Brunette, Register of Merit, 606 lbs. butter, 9873 lbs. milk.
Sybil's Oxford Princess, Reg. of Merit record, 605 lbs. butter, 8516 lbs. milk, 3 years 4 mos. old.
Oxford Spotless, Official Island test, 592.65 lbs. butter from 8848.50 lbs. of milk in 397 days.
Sybil's Oxfordia, Register of Merit test, 546 lbs. of butter from 6347.3 lbs. of milk in 310 days.

- Sybil's Gamboge was sold at auction in the USA in 1919 for the world record price of $65,000 – about £13,000! The group of financiers and Jersey cattle enthusiasts that bought him later paraded the bull down Wall Street!

- "Munifordia's Oxfordia 4th, the Island's highest lifetime production cow, with her breeder, Society Life Member Francis Le Ruez and his daughter Elisabeth. This great cow lived to 21 years of age and produced more than 109,000 litres of milk in her lifetime! Francis Le Ruez and his two brothers John and Laurence, are recognised as some of the greatest breeders of Jersey cattle and are part of a family that have been very influential in breeding Jerseys since the early days of the breed. Their grandfather was Francis Le Brocq, whose picture appears later in this book.

- Charles Mourant, in middle, with cup, and one of his high producing cows.

Jersey Cows

milk recording procedures. Yields were weighed on farm every two weeks, by milk inspectors who recorded each cow's milk at two daily milkings, taking small samples of the milk to the Society's laboratory where they were analysed for butterfat content. The yields on each test day were multiplied by the number of days between each test, eventually producing a full lactation of 305 days, or ten months, the milking period of a cow's annual production cycle.

From the officially recorded lactations of cows, comparisons could be made with others so the farmer could use the statistics in the selection of offspring for the succeeding generation. Individual cow's records were also used to evaluate the ability of their sire, and their dam, as progenitors of superior cattle.

- A dairymaid with butter-making equipment.

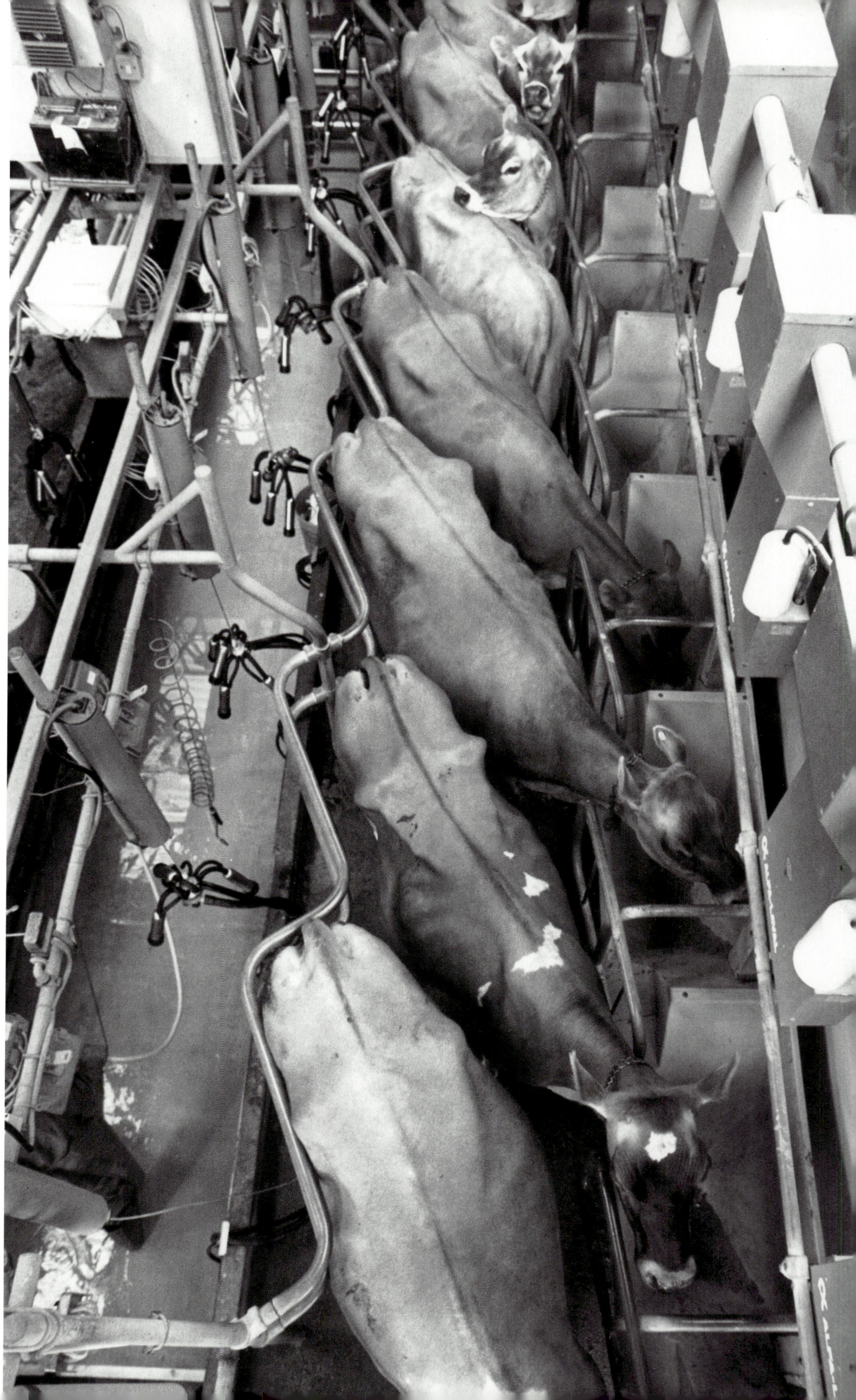

- An Unusual view of a modern 'herringbone-style' milking parlour. Fairview Farm, St Saviour, 1993

- The stable at Roselands in 1952.

- A Gascoigne Bail milking machine – the third machine to be installed in Jersey. It ran on a petrol motor until converted to electricity. John Orange Arthur of La Pompe, St Mary. (c.1938)

- The Gerber test for analysing milk samples at the milk recording lab of the RJA&HS, Mulcaster Street, in November 1939.

Jersey Cows

The cattle shows of Jersey have been the shop window for the breed since 1834, when, in contrast to the spectacular displays of rings of cows paraded in front of an eager audience, the cattle were judged "behind closed doors" individually by three judges.

The first show was held on Easter Monday, March 31st, 1834 in the Cattle Market, in Beresford Street. The appointed judges, Messrs. Bréhaut, Nicholas Le Bas and Edward Le Gresley, had to sign a declaration of impartiality, and it is on record that in one competition, the three gentlemen, having awarded equal points to two animals, had recourse to tossing a coin to decide which should be awarded the prize!

The farmers, having transported their cattle to the Cattle Market before eight o'clock in the morning, had to withdraw before the judging began. The cattle were then arranged by parishes, the parochial arrangement being kept secret and unknown to the judges. A special committee recorded, in silence, the awards of the judges, and countersigned the signatures of said gentlemen. It was not until the judging of each class had been completed that the members of the Society were admitted to the Cattle Market to inspect the animals. The general public were also allowed to view the cattle, but at a price ... it cost them sixpence for admittance!

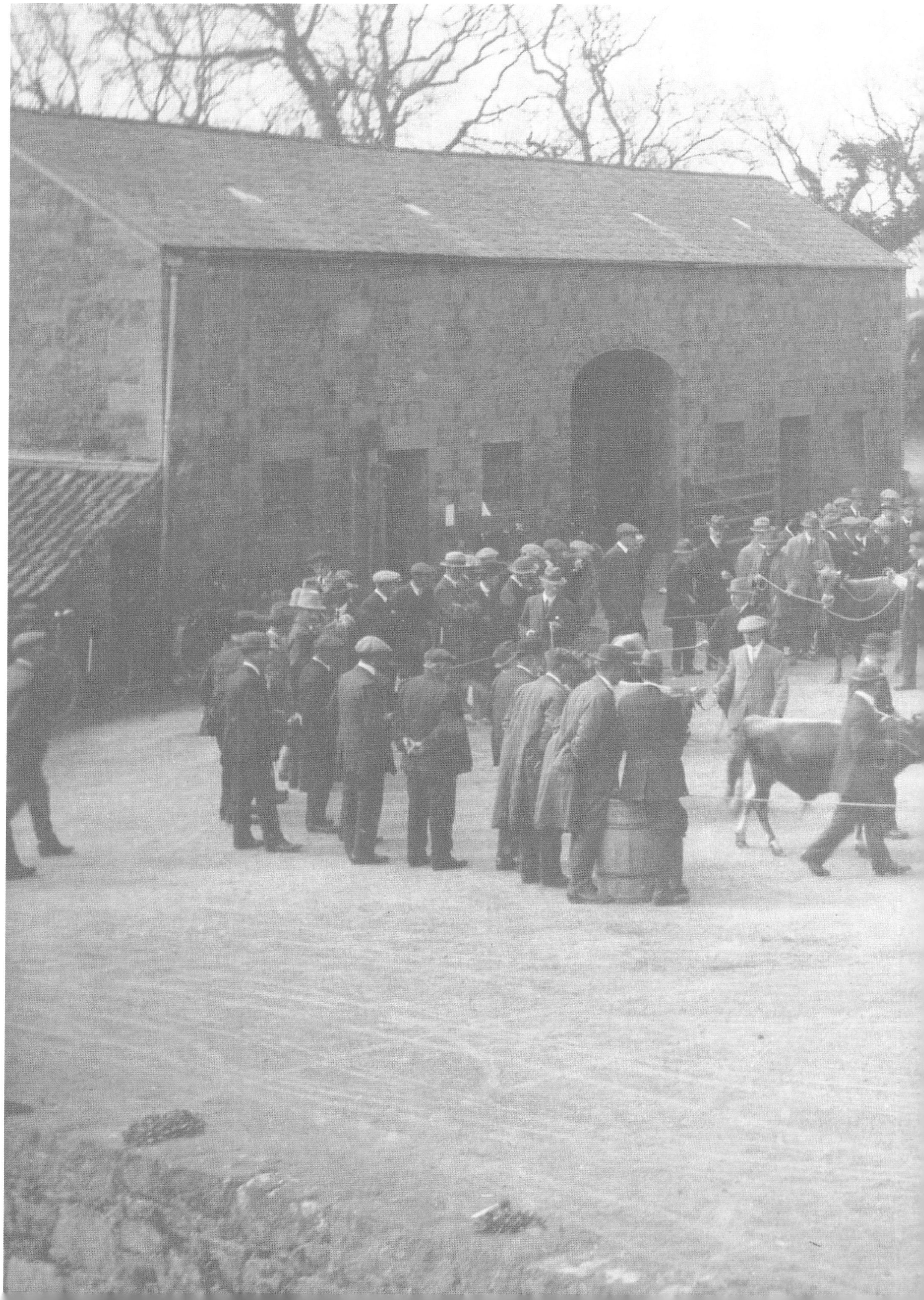

- The Western Three Parish Bull Show held at Les Niemes Farm, St Peter (c.1925)

- These pictures demonstrate the improvement in the form of Jersey cows in the past century and a half. The cow on the left was a good example of Jerseys in the mid 19th century, with a longer, plain head, a "humpy" top line and a thick tail-head. The udder was not well attached, and she had large teats. The cow on the right is a good example of the kind of cow farmers strive to breed today, with a good, dairy-like head, level top line, good udder attachments, both front and rear, and tects that are well placed under the udder and of uniform size, both important when the cow is going through a milking parlour.

Norman Walker ©

In those early days the first scale of points included four for the reputation of the animal's pedigree for producing "rich and yellow butter", and eight points which was nearly one quarter of the total, describing the head of the animal. Today, more emphasis is placed on the general dairyness, and in particular, the mammary, or udder, parts of the cow.

From the first show, colourful rosettes were awarded to prizewinners, along with a monetary award, and to win a red rosette for first prize "over Jersey" was to ensure a greater value for that animal. Blue, yellow and green followed, but the greatest prize of all was the red, white and blue rosette … for the champions!

The early shows were primarily the benchmark for attaining better conformation in the cattle, but as the breed improved rapidly in the mid-1800s, the shows took on a new meaning, with farmers competing with each other to demonstrate the qualities of their cattle, and to woo overseas breeders who regularly came to the Island to buy stock for their herds.

In 1871, the Channel Islands Exhibition, the greatest event in the social life of Jersey in the nineteenth century, was staged at Victoria College, and continued for three weeks. During

- The Island Spring Show, May 1946 and the first to be held at Springfield after the Occupation. Older breeders often referred to this particular show as the finest ever held.

- Judging the Island Spring Show in 1948. A wonderful line of mature cows, headed by Ernie Gaudin's "Design's November Nance". The judges, walking behind the class of cows before announcing their final line-up were Messrs A E Falle and J J Le Brocq.

- The Young Cow class, for three to six year old cows, at the Island Summer Show of 1925. Heading the class was Carlyle Le Gallais cow, "Septlaise". The judges, on the far left of picture, were Messrs J H Le Boutillier and C P Journeaux.

- The Island Summer Show of 1931 – these summer shows were two-day events, which included horse competitions and military pageants.

THE SHO

The spring of 1919 saw renewed interest in the Jersey breed following the First World War. The Island Spring show was held on May 21st, a beautiful spring day in Jersey and it was a typically glorious occasion for the farming community.

Visiting the Island at that time were soldiers from the USA, Canada, Australia and New Zealand, who had fought on the battlefields of Europe, but many were from farming backgrounds, and wanted to see the Jersey cow in her Island home, before returning to their own homes.

One man in particular, who visited Jersey that spring, was an American photographer, Harry Strohmeyer. He was the world's leading animal portrait photographer, but his duty at that time was to capture pictures of the war. This was to be his first of several visits to Jersey and he took the opportunity to tour all the parishes, capturing some fascinating pictures of the island at that time. The pictures on these two pages are a unique record of that Island Spring Show.

Harry Strohmeyer returned to Jersey in 1932 and again in the1960s. His pictures have been used throughout this book and allow the reader to have a glimpse of Jersey through the lens of an American visitor and enthusiast of the Jersey cow.

W OF 1919

this time, competitions were held for all kinds of animals, including 274 head of cattle, nine of which had travelled from Guernsey to compete in their own classes.

During the exhibition, 30,000 visitors attended with 10,000 visiting on the first day alone to see the magnificent show of cattle, and the large range of agricultural implements in addition to all the other displays.

During the nineteenth century, each of Jersey's twelve parishes established its own agricultural society, the first one being St Peter, who held their first show in 1846. The other parishes soon followed, some of them combining to hold joint shows with each individual parish having their own competitions in the morning to come together for the "sweepstakes" – for the best of the combined show - in the afternoon. These parochial events preceded the Island shows by around two weeks and provided an interesting guide to what might be expected when the parishes came "head to head" at the subsequent all-Island event.

The Island shows were held three times annually, in May, August, and October. In the early years, these shows were held in various

- Country politics – at St Lawrence and St John Parish Show at H.P. Vibert's farm, Midlands, May 16th, 1955. (left to right) Gerald Richardson, Gordon Picot (standing), Ernie Gaudin, Bill Perchard, Thomas Le Marinel, John Hidrio, Charlie Le Couteur, Jack Hidrio and Philip Binet.

- St Clement's Parish Show, May 1939. This is where the Parish Hall is now situated.

- St Brelade's Parish Show, May 1939 at the old racecourse at Quennevais.

- Waiting for the class in the cattle lines.

Jersey Cows

locations, usually the cattle markets, but sometimes at the People's Park, until in 1883, the Jubilee year of the Royal Jersey Agricultural & Horticultural Society, a committee was named to select a likely site for a permanent show yard. The need for this was obvious when, a year later, the Bull Show showyard extended to the Cattle and French Markets, the Prince of Wales Rooms and the Britannia Stables!

The Society purchased fields on the northeast of the town of St Helier, now known as Springfield, and a grand inauguration show was held in August 1885. Springfield was the venue for Island cattle shows for 110 years, until the Society sold the showground to the States of Jersey in 1994.

Queen Victoria's Jubilee was celebrated by a combined spring and summer show held in June 1887, and was the first cattle show when the animals were judged in public – 43 years after the first show! This was quite an attraction for spectators, happy to pay 2/6 – a considerable sum at the time - for the privilege of admission!

Since then, the pattern of the shows has remained largely the same – but now with no admission charge - with the "ringside judges" enjoying the fun, not only trying to second-guess

- The Grand Parade of prize-winners at the Island Spring Show in 1949.

Jersey Cows

the officials in the middle of the ring, but also having the opportunity of saying just how wrong they thought the judges were!

In the halcyon days of the export business, it was not unusual for a farmer, leaving the parade ring, proudly displaying the rosette won by his cow, bull or heifer, to be approached by a representative of some overseas cattleman, to sell the favoured animal.

Trade was brisk, and the prices reflected this. The shows of Jersey were truly the shop window for the breed!

- The first Island Show after the Occupation 1945 - note the British flag proudly flying!

- A town boy waiting for fresh Jersey milk! Springfield, May 1939.

Cattle Shows

- The World Jersey Conference Show, Springfield, May 1979.

- A shy young Rosemary Ahier (neé Perchard) with daddy's Champion cow.

- "Going home"

Jersey Cows

Like all forms of competition, exhibiting cattle does not just happen. It requires expertise in recognising which animals to show, and to prepare these animals so that they are in the peak of condition for the competition.

The art of showing pedigree cattle starts with the talent of the breeder himself, and his ability to carefully breed animals that will – with a slice of luck - eventually become champions.

The farmer studies his cattle and selects those, which, in his opinion, are worthy competitors, weeks before the show. The chosen animals are then put through a regime that will gradually bring them to a peak of perfection for the judges' eyes. This starts with covering the animals with a thin hessian sacking coat that acts as a continual brush as it removes the old hair from the animal's coat leaving a sleek, shiny one. Training the animal to walk may seem unnecessary but there is a lot of difference between it ambling along grazing and parading like a prima donna, with the head held high!

A week or so before the show, the farmer gives his protégé an overall shampoo which helps the coat to shine. Then comes the more exacting task of clipping, taking off the excess hair from the animal's head, neck, shoulder, tail, and legs. If the animal is a milking cow then her udder is carefully clipped

- "All hands on deck"

- left to right:
  "Clipping"
  "Plaiting the tail"
  "Wrapping it up"
  "Just a 'tot' before the class!"

• "Great expectations"

• "Spit and polish"

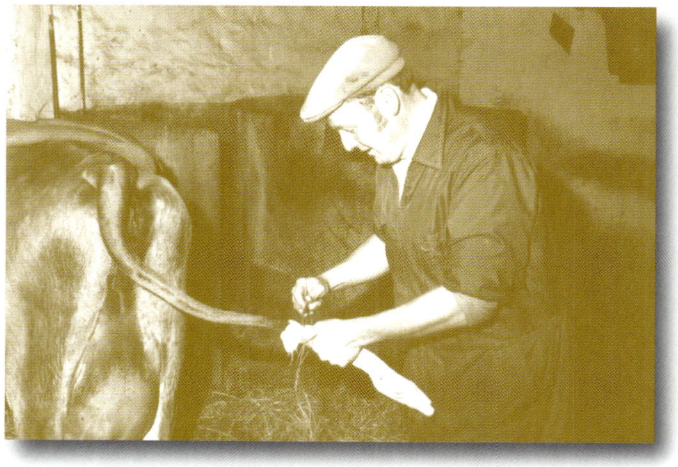

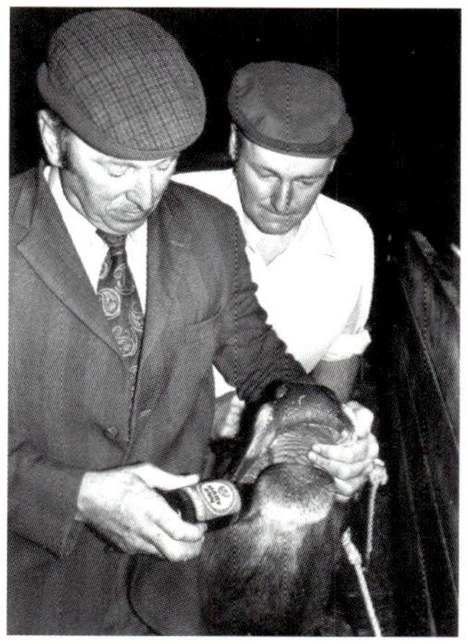

with fine hair clippers so that her milk veins stand out, to accentuate her dairy qualities.

The work of an expert clipper greatly enhances an animal while poor clipping can do the opposite. If an animal has horns, and few today remain, the farmer will spend hours rasping, sanding, and finally oiling them to a shining glory. In the past, when all cattle were tethered and their horns were a vital part of their control, farmers started training those horns from a very young age by gradually curving them inwards, making small niches on the inside of the curve, sometimes using a wire which would gradually be tightened as the horns curved towards one another. Special horn trainers and small weights that slipped over the tips of the horns were also used.

The evening before the show is always a hive of activity on the farm, with final preparations for the big day. The tails, usually the last part of the beauty treatment, are washed with soap and water, before being plaited, usually in three parts, and wrapped in rags and string so the animal does not soil them before the show.

Lactating cows are milked at variable times; depending on the amount of milk they are currently producing, and the timing of their class at the show. The farmer needs to show his cows with their udders at their best, and this is a particular skill which experience alone dictates.

- "Aspirations of the big prize!"

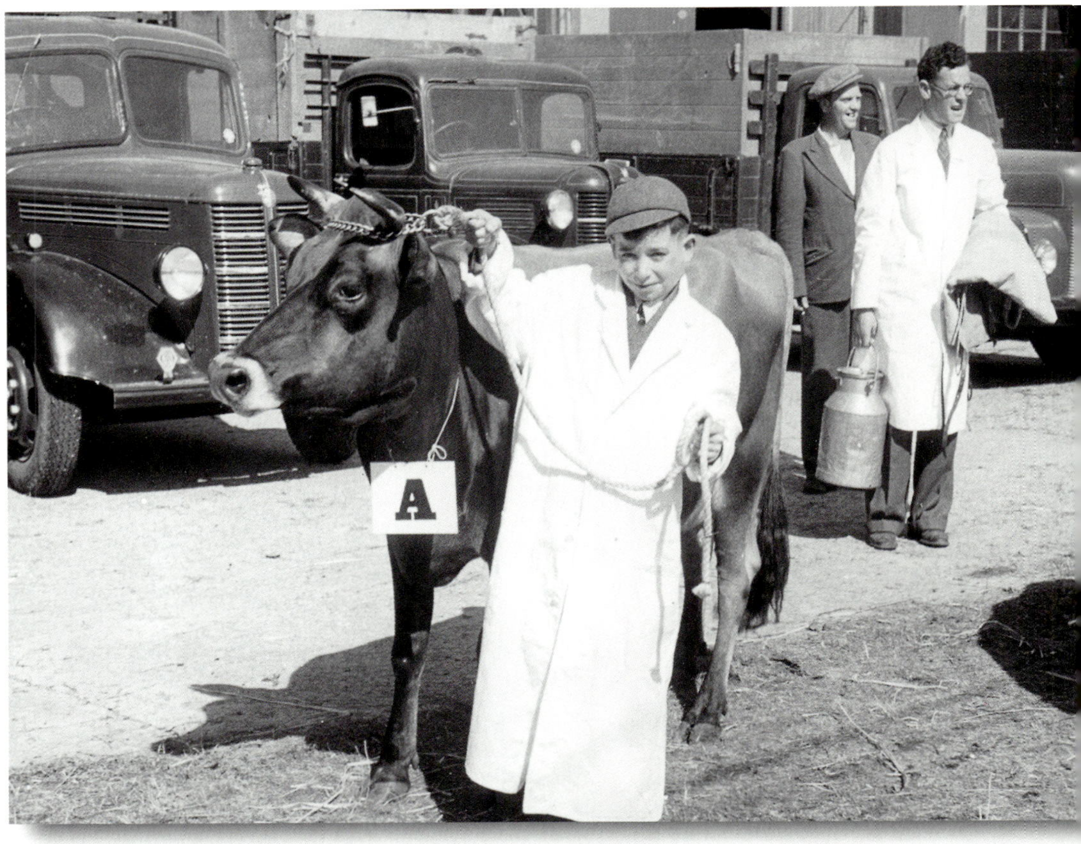

- Showing at the Skyline – indoors in October 1987.

• A busy scene in the Springfield barn.

• A progeny group – five outstanding daughters of the sire, Dreaming Victor (c.1952)

• The author learns the ropes at an early age (c.1949)

Jersey Cows

Having spent many hours in preparing each animal, the farmer proudly arrives at the showground with his team for the day's competitions. The first task is to settle the animals in the show barn, or, at parish shows, along a section of the specially erected holding ropes. He feeds and waters them and starts the final preparations – washing the animals, once more stripping the tails of the rags and brushing them out to complete the traditional beauty treatment.

Finally, just before entering the ring, soft brushing of the animal's coat, a little oil on the horns and feet, and a well-polished brass and leather halter, or if the animal has horns, a chain, completes the work. The bovine beauty queen, or king, is ready to compete!

- Clifford and John Pallot of St Martin with Bouquet's Cheerful Zebre, May 1972. "Cheerful Zebre" was Island Supreme Champion on two occasions.

- Gerald and Doreen Richardson of St Martin with Lynn's Dazzling Lady, August 1975. The "Lynn - Lady" family have had more Island Supreme Champions than any other cow family.

- Royal Dainty Maisie, bred by Snowdon Le Cornu of St Lawrence – the only cow to win Supreme Champion over Jersey five times!

- Ernie Gaudin of St Clement leads Dreaming Samudra at the head of the grand parade, August 1979. "Samudra" was Supreme Island Champion twice and lived to nearly 25 years of age!

- Stan Le Cornu of Trinity with Skippy's Devout Girl 2nd (left) and Biscay's Brownie (right). He scored a unique double by winning both Supreme and Junior Championships at two successive Island shows, in August and October 1970.

Jersey Cows

Jersey cows have been global travellers for many years and there is plenty of evidence of their travels to far-distant shores since the 17th century. It is important to reflect why the breed, originating in this little island, is the second largest dairy breed in the world today.

Her early popularity amongst the nobility of England was due mainly to the rich milk she produced, ideally suited for butter making; but her beauty also struck accord with the gentry. She was a proud status symbol grazing the parklands of grand houses in days long ago.

As those capabilities for producing better quality milk became known, her popularity increased and more and more landowners and farmers in Britain sought the thrifty little Jersey. Because of her size and ability to produce milk with less feed than larger cattle, she was also an ideal "ship cow" and travelled to many parts of the world, providing milk for the ship's crew and passengers along the way.

The settling of Britain's colonies contributed to the spread of the breed, especially in Africa and Australasia, and the Americas have had Jerseys for several hundreds of years.

Records show that a few Jerseys arrived in America as ship's cows as early as 1657. It is probable that the island's seafarers who sailed to the cod fishing

- Mr and Mrs Francis Le Brocq. He bought and shipped over 30,000 cattle in his lifetime, all in the days before the motorcar!

- "The Cornelius Grinnell", which landed 33 Jersey cattle in New York on 10th July, 1869

- Khedive's Primrose – this cow sold for the phenomenal sum of £1,000 in 1882! This would have enabled the purchase of a small farm at that time!

- Jørgen Larsen (far right) from Denmark negotiating the purchase of a heifer in 1906. Denmark imported 5,291 cattle from Jersey between 1896 and 1909, to establish their national herd that today is famous throughout the world. Mr Larsen founded the Danish Jersey Cattle Society in 1902.

- A shipment of cattle to Denmark. Each animal was individually loaded from the quay to the hold of the boat. (c.1898)

areas of the western Atlantic took cows with them. A Captain Pratt, master of the ship Hudson had taken many Jerseys from the Island to New York before anything was known about the breed in that country. He traded at ports on the coast of Africa, sometimes calling at Jersey on his return home. When in New York he amused the merchants with his sea yarns, telling them of the wonderful cattle he found on Jersey Island, and some of them, interested, asked him to bring back two or three with him on his next voyage. In is recorded that Captain Pratt frequently took Jerseys to America in the mid to late nineteenth century.

The American Jersey Cattle Club, the official breed association, was established in 1868 and proved a great stimulus for the breed with increasing numbers of cattle being exported from the Island. By the end of World War II more than 27,000 cattle had been exported from the Island to the USA and Canada, and they were the sole progenitors of over two million animals registered in the American Jersey Herd Book at that time.

Imagine the journey for those four-legged island envoys in the days of sailing ships. Put on board at St Helier harbour it would be four to six weeks later when they would sight land again, following a rough sea journey, restricted movement and probably not too much food! Then the final indignity, at the end of the journey, it was most likely that the animals would be cast overboard to

• A group of bulls settle down below decks for their long journey across the Atlantic in the 1950s.

• Cattle at Roseland, St Saviour, about to leave the farm for the USA shortly after the Occupation.

- Hedley Maillard, Senior and Junior, auctioneers, at a dispersal sale at Les Cateaux, Trinity in 1938.

- The bull, Observer, being transported to the quay for shipment to the USA (c.1919). This bull was very influential in the development of the breed in North America, and most American Jerseys today carry his bloodlines.

swim ashore – what a way to treat such important ambassadors!

Wonderful stories abound of the travels of Jersey cattle. New Zealand first saw Jerseys as far back as 1862 and in 1876 it is recorded that a Jersey cow named "Jenny" was purchased for £40 by a Mr Hulke, who then led her by a rope for a distance of 130 miles to Taranaki, where she became the Champion cow of Taranaki and also a winner of a milking contest.

The first record of Jerseys being imported into Australia was in 1854, and until 1888, Jerseys were imported into the state of Victoria, usually as "ship cows". Jersey cows were placed aboard the ship before leaving England, for the express purpose of supplying fresh milk to privileged passengers during the voyage to Australia. Interestingly, later importations included cattle for the Queensland Government, and their agent, a Mr Jones who visited Jersey to select animals went to great pains to choose cattle from the "highlands" of Jersey – the northern part of the Island, because, in his opinion, the ones from the lowlands (the south) would not stand up to the severity of a foreign climate as well as those bred at a higher altitude.

Cattle leaving Jersey for Australia and New Zealand spent many months on the high seas. On arrival, local farmers went aboard ship to inspect each

- In 1930, everything was done by mail! Farmers were reminded of their obligations by card in those days.

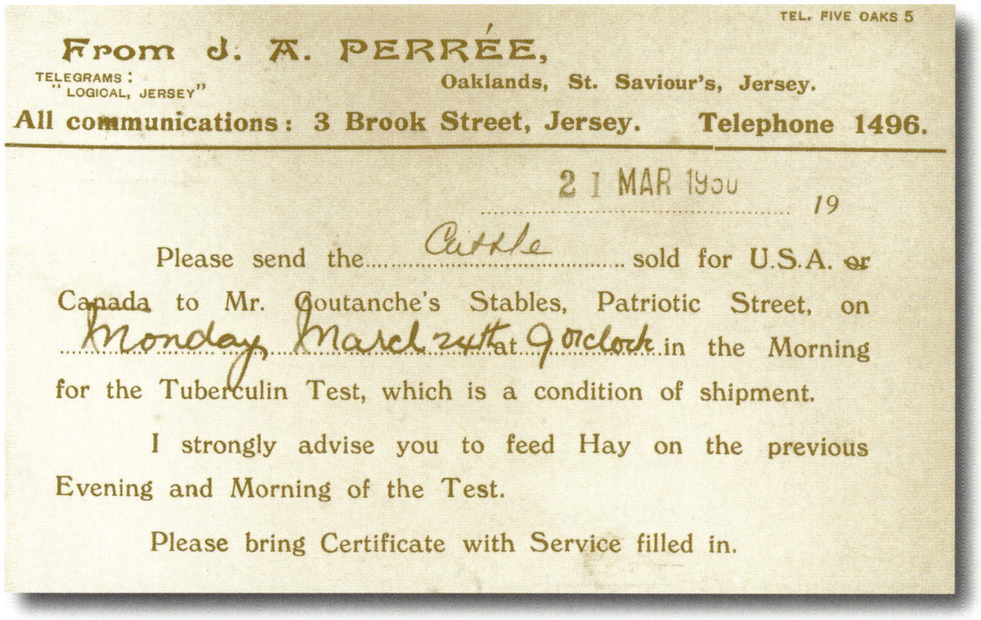

TEL. FIVE OAKS 5

From J. A. PERRÉE,

TELEGRAMS: "LOGICAL, JERSEY" Oaklands, St. Saviour's, Jersey.

All communications: 3 Brook Street, Jersey. Telephone 1496.

21 MAR 1930 19

Please send the Cattle sold for U.S.A. ~~or~~ Canada to Mr. Goutanche's Stables, Patriotic Street, on Monday March 24th at 9 o'clock in the Morning for the Tuberculin Test, which is a condition of shipment.

I strongly advise you to feed Hay on the previous Evening and Morning of the Test.

Please bring Certificate with Service filled in.

- Another shipment of cattle ready for their boat journey. The popular "bull-nosed" Commer lorry of the 1950s was a typical mode of transport. In those days, farmers purchased a lorry before they bought a car, and lorries were used for working as well as for family outings.

- Cattle off to Canada by Bristol Freighter in December 1955.

- A Pan-American Airways DC6 landed in Jersey in April 1959 to take this consignment to the USA. It was the largest aircraft to land in the island at that time and work to extend the runway by 300 feet to accommodate it had been completed just weeks before.

PAN AM

Jersey Cows

animal that was brought on deck for individual appraisal and would bid for their favourites at a subsequent auction.

South Africa may well have had Jerseys as "ship cows" in the mid nineteenth century, but the first documented records show that a Mr Adrian van der Byl from Cape Province bought cattle from the Island around 1877.
The Scandinavian countries have become home for millions of Jerseys, descending from the original importations, firstly to Sweden in 1893, and to Denmark, in large numbers from 1896 to 1909. The story goes that the Americans bought the "fancy" cattle on the Island, the English bought the best production ones, and the Danes bought "the ugly ducklings". True, the Danes paid very modest prices for their selections, usually from £10 to £15, but were valuable customers and took several thousand heifers to start what has become one of the world's outstanding national Jersey herds.

Latin American countries abound with Jerseys. Costa Rica was an early destination for the breed, in 1873, while Brazil and Guatemala had their first Jerseys in 1896.

The Jersey cow is truly the island's number one ambassador. Following World War II, the Jersey was the world's most popular breed with over two million farms worldwide being home to descendants of cattle from the Island.

- Exports of live animals from the Island are less common these days, as the costs of freight have risen dramatically. Today, the Island's best bloodlines are exported through frozen semen, used to breed cattle all over the world. Here, individual 'straws' of semen are prepared at the A.I.Centre laboratory.

- Loading on this special occasion was attended by the Island's Bailiff, then Sir Alexander Coutanche, later Lord Coutanche of St. Brelade's & Westminister and the Lieutenant Governor Sir Edward Grassett

• Gathering at the docks in front of warehouses that now house the Jersey Maritime Museum.

• Jersey sale catalogues from the turn of the 19th–20th century were works of art in themselves.

• In the 1950s and 60s, annual promotional Shows and Sales were held with many of the animals destined for export to the United Kingdom. In the ring leading their bull were Horace Ahier and his son Jeff, of Trinity.

Jersey Cows

Traditionally most farmers in the Island kept a few cows as part of a mixed farming operation that included cattle, sheep, pigs, apples, potatoes and fishing. Farmers took their milk and butter into market for sale along with other farm produce.

A number of dairy companies were established around the Island, each with their own collection rounds and retail outlets. In the early part of the 20th century there were 30 private dairies collecting milk from farms and selling directly to the public.

In 1952 the States Committee of Agriculture provided enabling legislation through a Marketing Law to establish the Jersey Milk Marketing Board in 1954.

The JMMB operated its own dairy to deal with milk surplus to the retail supply alongside the private dairies for a number of years, but eventually the private dairies closed and the JMMB assumed responsibility for the collection, processing and marketing of all milk produced in the Island. In 1966 the new dairy was built at Five Oaks to centralise the whole operation as the only milk processing plant in the Island.

The JMMB is a farmers co-operative but as the size and scope of the business expanded, responsibility for all commercial matters relating to the running of the dairy and the marketing of milk and products was vested in the Commercial Executive, already

- Delivering dairy products the old-fashioned way – by mule! This was one of the exhibits at the Society's Centenary Pageant in 1933.

- Fairview Farm Dairy of Eugene Perrédès at St Saviour, exhibiting at the Island Autumn Show of 1936. Note the signs advertising people to "drink milk from Mourma" – the Champion cow of the Show. The cow in front of the display is Mourma with Miss Adele Perrédès. For many years, the Fairview herd was by far the largest herd in the Island.

- A scene that has virtually disappeared – milking by hand in the old stable.

- A farm dairy complete with butter churning equipment (c.1932)

members of the Board of the co-operative. They adopted the trading name of 'Jersey Milk'.

As for the structure of farms, in 1930 there were over 12,000 head of cattle on 1,808 holdings, whereas in 2001 there were approximately 7,000 head of cattle on 47 dairy farms sending milk to the dairy. Like most agriculturally developed countries, Jersey now has a quota or licence system to regulate milk production at a level that can be marketed by the Dairy.

The developments in farm structure, driven by economic pressures, reflect the changes that have occurred in the agricultural industries in all developed countries. Herds have become larger and the yields of milk per cow greater with only the most efficient milk producers surviving.

- Unloading milk at the butter-making dairy in 1939.

- Heavy falls of snow meant milk collections by tractor in 1979 – this one opposite St Ouen's Parish Hall.

- A typical JMMB 5 gallon aluminium milk churn used in Jersey prior to bulk tank collection.

• A JMMB delivery truck.

• The dairy at Val Poucin Farm, Grouville in the 1960s. This 'model' farm was owned by the Rt.Hon.The Earl of Jersey who was President and later Patron of the World Jersey Cattle Bureau.

part 5

Harvest

Harvest

It has been many centuries since Jersey farmers considered cereal growing as a major part of their farming, but along with the other crops, wheat and oats were grown, on a small scale, to provide bread for the family and to feed the animals.

In Feudal times, wheat was the main crop of the Island and the staple food. It was used as currency, and the tenants paid "rentes" by supplying a portion of their corn to the Seigneur, the local "Lord of the Manor". Until 1797 the amount due was assessed on the current wheat price, even if the debt was paid in money rather than the corn. The price was eventually fixed by law and did not vary with the seasons.

Even in the 20th century, mortgages were still valued on paper as "froment de rente" or wheat rent. Today, parish rates are still based on "quartiers", or quarters, which stems from the feudal system of paying with a measure of corn.

The early farmers harvested their crops using scythes and sickles, and then separated the grain from the straw by treading the heads of grain or beating with flails, systems still used today in many developing countries.

Following the flailing, or threshing, the husks, chaff and straw remnants were sieved and blown. Known as winnowing, this left the grain ready for milling.

In the mid-1800s, the introduction of reapers and binders drawn by oxen or horses replaced the scythes

- Drilling corn seed in the 1950s. Philip Mourant of Mont au Pretre, St Helier.

- Cutting oats at Les Niemes, St Peter, the Le Rossignol family in the 1920s.

- Cutting hay at Roselands, St Saviour in 1932. This shows the density of growth and the excellent crop.

- Cutting hay with scythes in 1934. In the middle was Douglas de la Haye of West Lynn, St Ouen, with two farm workers, father and son Le Lai.

- Cutting oats with a sickle in the 19th century.

and sickles. The earliest mechanical thresher was patented earlier in 1734 to be succeeded by a more successful version in 1778. The power source, before the steam engine, was the treadmill operated by men or animals.

Further developments in the 1800s led to the box-like machine familiar to visitors to threshing fayres. The thresher takes in the sheaves of corn on the top floor, separates grain from straw, delivers the chaff and short straw at points around the machine and delivers the grain into sacks ready for the miller.

By then, power was provided by the steam engine with its flywheel and belt driving the unit that provided the final link in the mechanisation of the harvest work.

In Jersey, the annual harvest and threshing was very much a community gathering. Farmers grouped together to help each other collect the stooks of cereals; in the earlier days each one taking his horse and van to the fields – later it was lorries, or tractors and trailers; and then on to the local threshing machine for the days work.

They were busy days for everyone, and the farmers' wives provided baskets of sandwiches and Jersey wonders that were washed down with cider – threshing was thirsty work!

In Jersey in the last century, Le Marquand Bros. of West Park bought nearly all the grain. They milled and used it in animal feeds that were sold back to the

• Farrell's Threshing Outfit at St Ouen in 1912.

• The Richard family's Threshing team at St John in October 1930.

Corn and Grass

- A threshing scene at Brampton, St Ouen, probably around the turn of the century.

- Corn stacks in 1940.

- Loading hay for winter forage.

farmers as animal concentrates.

These days, a combine harvester does all the jobs in one go, with much less labour required. Little combining is done in Jersey nowadays, but maize is grown, often following the potato crop, for winter cattle feed. The maize, or corn as it is known in some countries, is a common sight around Jersey's countryside in the summer and autumn months.

These developments in the story of corn in the Island are also reflected in growing grass for hay to feed the Jersey herd. During the spring and summer months Jersey cows produce their rich, creamy milk from the grass in the fields, however, winter feed has changed from hay, along with root crops like parsnips and mangolds, to a mixed diet based on grass or maize silage.

Traditionally, hay was made during the warm summer days to form the bulk of winter feed for the farm livestock, however, the moist sea air and changeable climate in the Island does not favour haymaking, which requires longer periods of dry weather to cure the crop. Today, most dairy farmers turn the excess spring grass growth into silage, a process also used with the maize, as it is less dependant upon the weather and provides more consistent forage.

The grass is cut and cured for a couple of days, a process known as wilting, prior to being collected and chopped into bite size lengths by a 'precision-

- An early combine harvester drawn by a Massey Ferguson tractor.

- Threshing at La Grande Maison, St John. The tractor and machine were models "Société Française"

- Cutting and baling hay in the early 1950s. Note the field of tomatoes behind.

• Building a straw stack at Len Pipon's, St Lawrence (at top of stack with straw boater)

chop' harvester. It is then blown into bin-like trailers that follow the harvester. These are driven off to the farm where the chopped grass is tipped onto a bunker, or clamp, where the heap is pressed down with a heavy tractor. When the clamp is completely full, the grass is covered with plastic sheeting and held down with old rubber tyres. It is imperative that air is not allowed into the clamp so that the grass ferments and gradually becomes 'pickled', forming silage – making for a good tasty meal for the cattle in the winter months!

- Turning grass, aiding the wilting process, ready for the precision-chop harvester

- Silage making in the clamp at La Ferme, St Martin.

- Blowing the precision-chopped grass into the accompanying trailer ready to take to the silage clamp.

- A scene "tedering" hay – Ler Pipon of St Lawrence in 1978

- Wrapping big bale silage at La Ferme, St Martin.

The Occupation

Main picture, centre:

'War & Peace', a gentle Jersey cow standing with Rex Ross, a breeder and importer of Jersey cattle from Oregon in the USA, in front of wartime relics, 1946

Smaller pictures, clockwise from top left:

1. Potatoes in the Park – every square inch of land was used to grow food, here potatoes growing in the People's Park, St. Helier

2. Potato Clamps, potatoes buried in mounds of earth to preserve them for eating later.

3. 'Blockage', potato lorries parked up waiting to load as shipping became erratic a week or so before the occupation began, June 1940.

4. A shortage of wheat flour meant that bread was made from potato flour instead.

5. Filling sand bags at Springfield

6. Lack of petrol meant that the vet was forced into doing his rounds the old fashioned way

102

The Occupation

Winston Churchill was presented with a cow from the Island in grateful thanks for his leadership through those dark days. Pictured is the presentation being made to his daughter, Mary Churchill, by the President of the RJA&HS, Carlyle Le Gallais, and the breeder Mr. A.J. Mourant at the Dairy Show in Reading, 1946.

Field Marshall Montgomery discussing tactics with Jack A'Court on a visit to the Island after the war.

WESTERN 1617.

28, HYDE PARK GATE,

LONDON, S.W. 7.

June 23, 1947.

Dear Mr. Legallais,

You will be glad to know that MAYBELLE OF THE ISLES has very successfully passed through this difficult winter, and is in extremely good health. Mr. Bruford, the Veterinary Surgeon, says:

"During the recent tuberculin test on your herd at Chartwell, I examined MAYBELLE thoroughly and I am pleased to report that she is in fine condition and milking well, and will I feel sure amply justify her selection as the best Jersey Cow on the Island last summer."

MAYBELLE's manners are charming, and sh is most affectionate. Perhaps you would care t convey this report to those who are the donors c this distinguished animal.

Yours sincerely,

Winston S. Church

Carlyle Legallais, Esq.
Royal Jersey Agricultural Society.

part 6

R.J.A. and H.S.

• Colonel Sir John Le Couteur (opposite)

R.J.A. & H.S.

The turn of the 18th and 19th centuries heralded the dawn of the Industrial Revolution, and realising that the proper cultivation of the land and the improvement of breeds of livestock in Britain was the basis of the nation's prosperity, Agricultural Societies were being formed to encourage good practice.

One of the first of these Societies was the Bath and West, formed in 1777 and during the next fifty years, many regional and national bodies developed. The Highland and Agricultural Society of Scotland was founded in 1784, but the Royal Agricultural Society of England was not established until 1838, five years after the Island's own Society.

- The Society has always played a unique role in the development of 'improved' farming. In 1900 this went as far as setting up a special committee to take an in-depth look at "the present state of our local agriculture". The two extracts set out here show a summarised conclusion of their labours.

The Committee have further applied the above test to the working of a farm with 45 vergees of land :—

YEARLY ESTIMATE of 45 Vergees of Land (Arable), with HOUSE & OFFICES not included in the measurement.

DISTRIBUTION OF 45 VERGEES—

Potatoes	20 Vergees	15 of which will be sown to roots and consumed by the Live Stock, and 5 to Clover Seeds.
Wheat	5 "	
Clover (1st year)	5 "	off which two cuts will be taken, the aftermath September to December for grazing.
Ditto (2nd year)	5 "	off which one cut must be taken, the aftermath July to December for grazing.
Pasture	9 "	
Vegetable Garden ..	1 "	
	45 Vergees	

LABOUR.—The farm to be worked by the farmer, his wife, one man and one woman servants; the raising of Potatoes and mowing of Hay being extra.

LIVE STOCK .. ..
2 Horses.
7 Cows,
3 Heifers in Calf.
4 Yearlings.
3 Calves,
Two sets of 6 Pigs during the year.

- Taken from the test example of a farm with 45 vergées of land, with a rental value of just some £4/vergée they suggested it would have been cultivated and managed by the farmer, his wife, a man and a woman servant. It is interesting to note that even in those days they anticipated bringing labour in for "the raising of potatoes and mowing of hay".

1833 marked an unprecedented transformation in the fortunes of farming in Jersey with the foundation of the Jersey Agricultural and Horticultural Society that, within a year, King George IV honoured with His Royal Patronage.

At a memorable meeting held on August 26th, in the Lieut.-Governor's Office, and presided over by His Excellency, Major General Thornton, C.B., the first resolutions were passed. Just twelve days later, a set of rules were adopted, along with the following resolution … "That it will conduce much to the general welfare of the Island of Jersey to encourage improvements in its agriculture and horticulture, the breed of cattle and to promote habits of industry and economy among the labouring classes".

The history of the Royal Jersey Agricultural and Horticultural Society, Jersey's oldest Society, reflects the development of farming in the Island in the last two centuries and much of the progress has been due to the guidance of the Society.

Five years after its inception, the Annual Report of 1838 recorded the following achievements…

"This Society set out with the design of creating a spirit of industry and emulation; it has fulfilled its object. It has improved, greatly improved, the breed of cattle. At the last show, one hundred and sixty head of very fine cattle, in a most superior condition as compared with former exhibitions, were declared by the Judges to evince in the most satisfactory and conclusive

- Today, an 'average' farmer is likely to employ just one full-time employee with seasonal staff when required. He would expect to pay a land rent of some £150/vergée, milk a herd of 100 cows and grow one crop - potatoes. His wife is likely to have found employment in the thriving local finance industry!

EXPENSES.

	£	s.	d.
45 Vergées of Land at £4	180	0	0
Labour.—Two men £78, 2 women £46 ..	124	0	0
" For raising Potatoes, 20 vgs. at 30s.	30	0	0
" For Cutting Hay	2	10	0
Expenses.—Threshing Wheat, 5 vergées at 8s.	2	0	0
Manure, Potato, 20 vgs., 600 lbs. per vergee, £8 per ton	48	0	0
" Potato, 1½ tons Bones at £5 10 ..	8	5	0
" For 5 vgs. of Hay at 10s. per vergee.	2	10	0
" " *5 " of Pasture " " ..	2	10	0
Food purchased for Stock	15	0	0
" " " Horses, 2 tons of Oats at £6 5s.	12	10	0
" " " Pigs, 12 months	11	15	0
Purchase of 12 young Pigs	6	0	0
Fodder bought, 2 tons Straw	7	0	0
Tithes on 5 vergees of wheat	0	17	6
Risk and depreciation on Live and Dead Stock, £385 at 6 p.c.	23	2	0
Sundry Expenses	10	0	0
Balance, being profit	45	0	6
	£531	0	0

* Of the 9 vergees, 4 vergees are manured with Liquid Manure.

RECEIPTS.

				£	s.	d.
20 vergees of Potatoes at £18 per vergee	£360	0	0			
Less value of 400 Cabots of Seed Potatoes stored, at 2s.	80	0	0			
				280	0	0
35 lbs. of Butter weekly at 1s. 2d., £2 0 10— 52 weeks				106	5	0
3 Heifers in calf, sold at £14				42	0	0
1 Yearling Heifer				6	0	0
3 Bull Calves at 20s.				3	0	0
12 Pigs of 100 lbs. each nett at 5¼d.				26	5	0
175 Cabots of Wheat sold, at 2s.				17	10	0
Profit on Vegetable Garden, estimated at ..				5	0	0
To the total Receipts must be added:—						
Increased value of 4 Heifers ..	£20	0	0			
" " of 3 Calves ..	15	0	0			
House accommodation, in lieu of rental	10	0	0			
				45	0	0
				£531	0	0

manner, the undeniable improvement that is manifested in form and condition.

Their value has also increased, thirty pounds having been refused for a prize cow, and her prize yearling heifer and twenty pounds have been refused for another prize heifer.

This Society has led to the adornment of cottages; fragrant flowers now decorate many of those yards, which, ten years since, were foul with manure and filth; thus, it has improved domestic economy and cleanliness, and therefore promoted health.

It has encouraged and rewarded long and faithful services. It has improved the general condition of many farms, hence increased produce.

It has led to the adoption of improved varieties of corn and increased its growth and produce.

It has proposed the cultivation of valuable sorts of potatoes in relation to commerce."

All these improvements were made in just the first five years of the Society!

In its formative years, the Committee conducted all the business of the Society, until 1864 when the two Departments, of Agriculture and Horticulture,

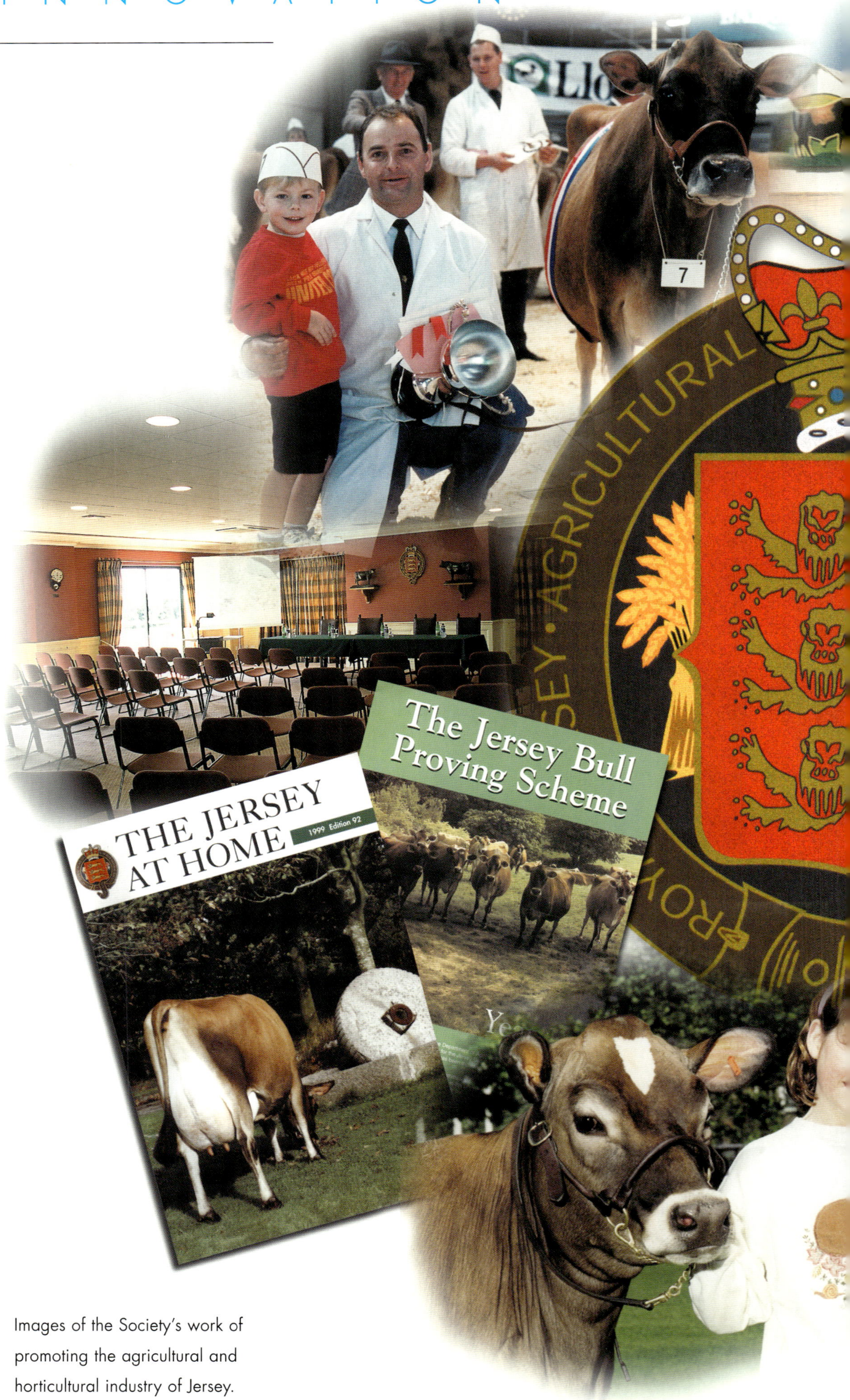

- Images of the Society's work of promoting the agricultural and horticultural industry of Jersey.

separated into two distinct and independent bodies but continued their incorporation under the common heading of the Society.

During its first century, the Society undertook virtually all the agricultural research in the Island, including crop trials and competitions to improve livestock and crops, and actively represented the interests of the farming industry. Both cattle and horticultural shows have been the stage for providing benchmarks in improvement. The Society had lobbied the States since 1902 to conduct experiments on the cultivation of potatoes and grasses, and in 1908, to celebrate its 75th anniversary, it made a strong representation to the States Assembly, which eventually resulted in the establishment of the States Farm Experimental Station.

The Society's long history has embraced much of Island life. Although its core work has always been that of an agricultural society, the RJA&HS has, for nearly 120 years, provided a venue for the public of Jersey to enjoy a variety of entertainment.

In the autumn of 1884, the Society obtained its Act of Incorporation that permitted it to purchase and hold real estate. A piece of land on the edge of the town of St Helier was bought by the Agricultural Department from Mr T.C. Le Gros. The property was known as Springfield.

R.J.A. & H.S.

The role of the Horticultural Department of the Society has always been less commercial than its agricultural counterpart, and in 1864, some conflict of interests grew and finally resulted in the two departments separating, each with its own autonomy but with the same Patron, Vice-Patrons and President. This situation has remained the same since.

- The Fruit Show of 1931. This was James Knott's collection of 104 varieties of apples grown at Samares Manor.

The Horticultural Department's business is that of organising Flower Shows and Garden Competitions, as well as encouraging better horticulture through debate and affiliation to most of the national specialist Societies.

Throughout its history, the Horticultural shows have been an expression of changing fashions; in 1852, for instance, Mr Saunders, a nurseryman, showed a class of 120 varieties of pears!

- A typical Horticultural Show of the 1930s set in the Springfield Hall.

In its formative years, long before the days of Garden Centres and the like, the Department was active in the encouragement of garden improvement around the Island. In 1872, it was responsible for a trial of lawnmowers, when "Green's Patent", exhibited by Messrs Le Masurier and Vibert, was found to the best.

But even in the demure days of the Victorian world, the Horticultural Department had to comment in the strongest terms about some entries at its shows, on the "reprehensible practice of exhibiting products

- A vegetable display arranged for effect at the Summer Flower Show of 2001, by Thomas Brown and Jordan Diver

- A wonderful display of flowers, fruit and vegetables in the 1930's

either bought or borrowed for the occasion" and issued a warning of stern measures in the future.

At one time, up to six shows each year were staged by the Department. The Imperial Hotel was often the venue in the mid-19th century, but the Victoria Rink, in Belmont Road, in the heady days of popular roller-skating, was also used for the gentler art of flower and vegetable showing. In 1880, the first of many shows was held at Government House, the home of the Island's Lieutenant Governors, many of whom have been strong supporters of the horticultural competitions. Major General James Frederick Love, Lieutenant Governor, in 1856 at the annual dinner, produced samples of tobacco grown and cured at Government House.

From the early days of the Society, many changes have taken place in the gardens of Jersey, and in the competitions linked to gardening. Dahlias and Chrysanthemums were in their infancy at that time, old varieties of roses were common, and heartsease grew wild in the native woods. Summer bedding plants were unknown and the beautifully laid out gardens we know today were practically non-existent, except perhaps in the large stately homes of the Island.

Today, the Horticultural Department holds three excellent shows each year, two in Howard Davis Park, and the other at the Society's showgrounds at Trinity. A strongly competed Garden Competition held each summer encourages many Islanders to work hard to keep the Island in a harmonious state.

- New competitors, Lester and Theresa Rondel, won the class for two to five perch in the 2000 garden competition

- Perennial garden competition winners Graeme and Janet Cloarec won their class in the 2000 event.

- Judges Max de la Haye and Arthur Ahier weigh this monster of a pumpkin.

R.J.A. & H.S.

The Society's first offices, in 1834, were at Mr Bernard Saunders' establishment, 19 Halkett Place, which remained the headquarters until 1865. The facilities included offices and a Board Room, and since that time until just a few years ago, the Society's offices were always referred to as "the Board Room". A move was made from the original "Board Room" to 19 Bath Street, and later, in 1899, to 8 Church Street.

In 1913, 70 years after its inception, the Society purchased its own office property, 3 Mulcaster Street. This prestigious granite building opposite St Helier's Church was formerly a flourishing hotel known as The Grasshopper. The Society remained there for 56 years, until 1969, when it was sold to a finance corporation.

The sale of the Mulcaster Street property financed the building of new offices at Springfield showground, and for the first time in its history, the Society's real estate was combined on one site.

The foresight of the Society's leaders in its Jubilee Year, 1883, to secure a permanent show yard, gave greater financial security for most of the next 120 years, but it was not always the case.

In 1902 the Society acquired a further piece of land adjoining Springfield. That was the Glebe land and it allowed the considerable expansion of the

- The opening of the "new" offices at Springfield by the His Excellency Lt.Governor, Air Chief Marshall Sir John Davis, on September 27th 1969.

- 3, Mulcaster Street in 1919. Standing outside were Wallace MacMonnies and Edmund Butler from the USA, visiting the Island to buy cattle with some local cattle breeders. The hotel next door was the Hotel de ''Europe (now Chambers) where they regularly enjoyed lunch and played dominos.

- Springfield Hall, built in 1924.

- Springfield was the venue for many public activities including the Battle of Flowers (1928-1939)

showground. The same year, a new entrance in Oxford Road was made, and sheds were erected against the outer walls on the northeast side of the property, thus creating the layout as known in the years before it was sold to the States. The showground was soon after used as a football pitch, which remains its main use today.

Springfield has been used for all kinds of activities but none more unusual than during the First World War, when it was commandeered by the Island's authorities and used as an artillery camp, a barracks and finally a food store. Handing back the property in 1919, the Society was paid compensation by the War Office. This income, along with the boom in the agricultural industry following the war, allowed the Society to build Springfield Hall that was officially opened in 1924.

At the outset of the project, the Society was in a good financial position, but a series of unforeseen events nearly plunged it into financial ruin. The problems associated with building, even in those days, were not insubstantial, and the great depression that loomed on the horizon did nothing to enhance the situation.

The cost of the building of Springfield Hall was £13,500, a small sum today, but in the 1920s, a very considerable amount. The Society was forced to raise debenture bonds to continue, and the tireless effort of a few determined committee members loyally backed by the general membership saw it through the stormy years.

One man, Sir Jesse Boot, a successful entrepreneur and founder of Boots the Chemists, was particularly helpful to the Society in its troubles. He lent a considerable sum of money, and when nearly all was lost, turned the loan into a gift, thus stabilising the Society's financial situation.

The outbreak of World War II and the five-year German Occupation of the Island interrupted the smooth running of the Society. The export of produce and livestock came to an abrupt halt, and cattle shows were cancelled. Springfield was again commandeered, but this time by foreign forces, for military use.

The early years following the war were again "boom" years for the farmers, and for the Society. Springfield again became the Island's primary entertainment venue, hosting every conceivable kind of event.

Since the 1920s, Springfield has been the home of

- The Officers of the Agricultural Department in 2001 alongside the Opening Stone. Left to right: James Godfrey, Chief Executive, Derrick Frigot, Senior Vice President, Lewis Rondel, President, Robert Perchard, Junior Vice President and Deputy Philip Ozouf, Treasurer

• Her Majesty The Queen signs the Society's Visitors' Book in the Members' Room. On the left is David Hambrook, Assistant Chief Executive and on the right is Society President Lewis Rondel, alongside the Bailiff of Jersey, Sir Philip Bailhache.

• The President welcomes Her Majesty The Queen and H.R.H. The Duke of Edinburgh to the Society's home, prior to Her Majesty unveiling the Opening Stone.

football in Jersey, and it has seen the likes of Manchester United and other top-flight teams grace its pitch. The Hall has also played its part in hosting some famous names ... try the Beatles, for one! In 1994, following considerable negotiations, Springfield was sold by the Society to the States of Jersey, and it is now, as it always has been, the premier sports stadium in the Island.

The sale of the property allowed the Society to move on to new pastures, literally, as part of the sale to the States included two small fields at Trinity, for the Society to build a new headquarters and showground.

Following protracted debate by the Society's members, the Royal Jersey Showground and headquarters was built and officially opened by Her Majesty the Queen on July 13th 2001, in front of Society members and a crowd of 3,000 people. The Royal Party toured the new facilities and viewed an exhibition of "Rural Jersey" celebrating the agricultural heritage of the Island. Only one thing was missing on that memorable day ... the Jersey cow. Restrictions due to the Foot and Mouth Disease outbreak in the U.K. prevented the Island's bovine ambassador from making an appearance!

• The Society's new headquarters in 2001.

R.J.A. & H.S.

When His Majesty King William IV conferred His Royal Patronage upon the new Jersey Agricultural and Horticultural Society in 1834, he set a precedent which has been graciously continued by each successive Monarch.

To celebrate its first year, the Society sent a gift of fruit to His Majesty the King. It included 103 Chaumontel, 36 Cressanne, 35 Beurre d'Arembert, 24 Golden Pippen, 48 Pigeonnet, 24 Waterford as well as other varieties of pears and apples, many of which have long since gone out of cultivation.

- The visit of His Majesty King V and Queen Mary in 1921. At Springfield, they enjoyed a display of prize-winning cattle.

In 1837 Her Majesty Queen Victoria ascended the Throne, and shortly after graciously granted Her Royal Patronage upon the Society. She visited the Island with the Prince Consort in 1846, when a gift of fruit was presented to her. The following year the Society sent a two-year-old heifer, purchased from Mr Thomas Filleul for £23, and a yearling bull with another heifer, given by the Society's President, Sir John Le Couteur, to Windsor, where Sir John, representing the Society, made a gift of the animals to Prince Albert. Today, the famous Windsor Jersey herd is still housed at the Prince Consort Farm in the Home Park at Windsor.

In 1887, Queen Victoria's Jubilee was celebrated by a show of the Society in June of that year, where the judging of cattle took place in public for the first time.

Queen Victoria, whose long reign nearly coincided

- The King being presented with the cow, La Sente's Miss Bronzemine, by the Society's President, Reverend G.P. Balleine. Rector of St Saviour. To his left is the breeder of the cow, Philip J. Bree of Grouville.

The Royal party leaving the showground.

with the life of the Society, passed away in January 1901. She was succeeded by her son, His Majesty King Edward VII, and in due course, He granted His Royal Patronage to the Society, although during his reign he neither visited the Island nor is there any record of a gift being made from the Society.

His Majesty King George V did, however, visit Jersey in 1921, where he was presented with a splendid specimen of the breed, "La Sente's Miss Bronzemine", at the Society's showyard at Springfield.

In 1936, King George V died, and His Majesty King Edward VIII took his place on the Throne, only to abdicate eleven months later. Four days after, His Majesty King George VI was proclaimed King, and following in the footsteps of His predecessors, including his brother Edward VIII, granted Royal Patronage to the Society.

King George VI and Queen Elizabeth, the present Queen Mother, visited Jersey less than a month after the Liberation of the Island from the German occupation, a visit that lifted the morale of the people enormously.

In February 1952, King George VI died and Her Majesty Queen Elizabeth II ascended the Throne, and on June 23rd, bestowed Her Royal Patronage upon the Society. In 1957, during a visit to the Island with Prince Philip, she was presented with the cow, "Beauchamp Oxford Lady" by the Society.

• Princess Margaret visited Jersey in 1959. While walking through the Roselands herd of John Le Gallais, one of the Jerseys took great interest in her floral dress.

• Her Majesty The Queen inspects the parade of cattle at Springfield in 1957. The Society's President, Francis Le Boutillier, accompanies her along the line of bulls.

- Her Majesty enjoys a conversation with Clifford Pallot and his Champion cow, Seer's Bouquet, as Francis Le Boutillier looks on.

- Society President, Francis Le Boutillier presents Her Majesty and Prince Philip with Beauchamp Oxford Lady, bred by Basil Le Brun, who is holding the cow.

R.J.A. & H.S.

In June 1978 the Society celebrated the occasion of the third visit of Her Majesty to the Island, as she had visited Jersey with H.R.H. the Duke of Edinburgh as Princess Elizabeth in June 1949, with the presentation of the cow, "Ansom Designette".

The latest visit of Her Majesty the Queen, and H.R.H. Prince Philip was on July 13th 2001, in the "Year of the Jersey" on the Island. During the visit, Her Majesty graciously performed the official opening of the Society's new headquarters and showground.

In his parting speech, the Bailiff of Jersey, Sir Philip Bailhache, promised Her Majesty that, when the Foot and Mouth outbreak stopped and normality returned, a Jersey cow would be selected and sent to Windsor for Her Majesty's fine herd.

The Society has been most fortunate in having Patrons who have an interest in breeding Jersey cattle. Her Majesty the Queen has a very successful herd of Jerseys at Windsor, which is one of the oldest herds in the United Kingdom. It is based on island cattle, blended with international bloodlines, and has been a leading source of good quality Jerseys for many generations.

- Anne Perchard, breeder of Ansom Designette, chats with Her Majesty at the display of country life held at Le Petit Catelet, St John in June 1978. Senator Bernard Binnington, President of the Committee of Agriculture accompanies Her Majesty.

- Society President, William "Bill" Perchard introduces Ansom Designette.

- Her Majesty is intrigued by La Grand' Tchéthue as Stan Morel, President of the Jersey Farmers' Union describes it's functions.

- John Le Ruez of Auvergne, St Peter talks about his cows to Her Majesty as Vice-President Frank Avrill listens attentively. Also on the left is Kate Le Ruez, John's youngest daughter. Her Majesty has purchased several outstanding cattle from this herd. (c.1978)

R.J.A. & H.S.

The Society's Centenary was celebrated in style at Springfield on August 23rd 1933 with a "Pageant of Island Agriculture".

The pageant was an expression of memories of the history of the Society and the agriculture and horticulture of Jersey displayed in a colourful and vivid way. The events were varied and included a grand parade of the previous day's winning show cattle, and a procession of every type of farm activity practised during the previous 100 years.

Leading off the grand parade was the Band of one of the premier cavalry regiments of the British Army, the 5th Inniskilling Dragoon Guards. Following them were forty superbly mounted horsemen in their scarlet and green uniforms with silver helmets. Later in the festivities, these horsemen delighted the crowds with a varied programme of Musical and Trick riding.

The occasion was enjoyed by thousands of islanders and culminated in a grand firework display ending with a huge flamed message "1833 – R.J.A. & H.S. – 1933"

- The 5th Inniskilling Dragoon Guards going through their paces.

- The display of farming "temps passé" included these gentlemen and a load of corn straw.

- The parade of exhibits going around the arena.

- Part of the mounted ride.

- These maidens are selling their wares – including the duck in the basket!

- The parade of prize-winning cattle.

- The band of the 5th Inniskilling Dragoon Guards.

part 7

A world of Jerseys

• Jerseys in California, U.S.A.

A world of Jerseys

The concept of creating a world organisation for Jersey cattle breeders stemmed from the thoughts of a South African farmer, Mr S D G Rossouw, a few years after the Second World War. At that time, dairy cattle, especially Jerseys, were travelling to all parts of the world, mainly from the Island and from England, to restock farms that had been depleted in the harrowing years of war.

There was concern about the quality of animals that were arriving in that country and being sold for high prices to young beginners. At a Special General Meeting of the Jersey Cattle Breeders Society of South Africa on the 16th March 1948, Mr G Vernon Crookes, their President, proposed "that the Royal Jersey Agricultural and Horticultural Society should be requested to convene an International Conference on Jersey Island in May 1949, where the classification, inspection, milk recording, import standards etc. for Jerseys could be discussed by delegates from all Jersey Breed Societies over the World to endeavour to bring about some uniformity and standardisation of the rules and requirements as regards Jersey cattle throughout the world".

The RJA&HS took up the suggestion and plans were made to invite delegates from countries with populations of Jersey cattle – the customers that had been so supportive of the Island and its cow.

- The group of international Jersey cattle breeders meeting at the Town Hall in May 1949.

- Lord Jersey officially opens the World Jersey Cattle Bureau' first office in 1992 with a large group of international Jersey breeders toasting the occasion in a glass of Jersey milk.

The Island's Bailiff, Sir Alexander Coutanche (later Lord Coutanche) opened the conference in St. Helier Town Hall on Monday 16th May 1949. During that week, a range of research papers were delivered by eminent gentlemen of that time; visits to farms and places of interest in the Island were included as well as the Island Spring Show.

On the last day of the conference, Mr E Lea Marsh, a delegate from the USA, suggested that the success of the week's events should prompt the formation of a permanent World Jersey Federation. The delegates took up this idea and Mr Albert Messervy, a local veterinary surgeon (he was later Emeritus Professor, and became President of the RJA&HS in 1975) was elected as Acting President.

Informal meetings took place during the following two years and in 1951, the first entry in the Minute Book recorded that "the initial meeting of the World Jersey Cattle Bureau took place at the Hotel Russell, in London, on Monday 21st October 1951."
The first Ordinary Meeting of the General Council was held at the same hotel on 4th July 1952, at which the Constitution and Rules were approved. Among other decisions taken, one was to accept the invitation of the Canadian Jersey Cattle Club to hold an International Conference in that country in July 1954. The World Jersey Cattle Bureau was on its way!

Since those days of fifty years ago, when ten countries were elected as the Founder Members of the organisation, the WJCB has grown to twenty-eight countries representing many thousands of Jersey cattle breeders worldwide.

The WJCB publishes journals and research information in all mediums, as well as helping young people to travel and work on Jersey dairy farms in other countries. Its offices are in the new RJA&HS headquarters where it shares the most comprehensive library of Jersey cattle books and information in the world.

The first President of the WJCB was a Jerseyman, the late Professor Albert Messervy, and in the Jubilee Year of the organisation, it is fitting that the President is also a native of this Island, Mrs Anne Perchard, M.B.E.

In June 2002, the 16th International Conference of the Bureau will be held in Denmark. This will be the latest in a series of global visits and meetings in more than thirty countries, where Jersey cattle breeders have got together in friendship and to learn from each other about breed and dairy management improvement.

A world of Jerseys

World Jersey Cattle Bureau

USA

One of the most important sources of Jerseys today, the USA registers more than 60,000 heifer calves annually. Canada too is a major exporter of Jersey cattle breeding all over the world.

Brazil

Brazil imported their first Jerseys in 1896, and the breed is growing rapidly today. The potential for Jerseys is enormous in Latin America's largest country.

Mexico

Virtually all of the Latin American countries have large populations of Jerseys – these Jerseys in Mexico are farmed in modern facilities that blend with traditional feeding methods.

Uruguay

The pampas of Uruguay is home to these Jerseys – Argentina also has Jerseys where the breed is extended throughout the varied climatic zones of the country.

Germany
The Jersey does well in Northern Europe and is particularly popular in Denmark where they are noted for their very high quality milk. These cows in Germany are part of a growing national herd of the Jersey breed.

Japan
Japan, where the breed was established following the Second World War, has a growing number of Jerseys. Admired for their high quality milk, this milk is used for a variety of different products.

India
India has more cattle than any other country, and the Jersey breed has been used extensively in cross-breeding with the Zebu cattle to improve production.

Africa
This Kenyan woman is one of thousands who have just a few milk cows on a smallholding. The Jersey is particularly suited to the climatic conditions of African countries.

New Zealand
New Zealand has one of the largest populations of Jerseys in the world with some 700,000 pure-bred Jerseys and another 750,000 cross-bred ones. These Jerseys provide much of the world-famous New Zealand butter that is exported to all parts.

Australia
Jerseys in Australia are mostly kept on a pasture system. There are more than a quarter of a million Jerseys in that country of so many environmental contrasts. The Jersey performs well in all of Australia's states.

A world of Jerseys

Gracing the 'Wests Centre' square in St Helier, just a few yards from where the very first show of Jersey cattle was held in 1834, is a beautiful and noble bronze sculpture of a group of Jersey cattle surrounding a drinking trough. This sculpture was the inspiration of Mr Aidan Smith, O.B.E. who, while travelling in the Andes Mountains of Peru, came across, unexpectedly, a herd of Jersey cattle. This set him thinking about the Jersey cow and all that it meant to populations throughout the world.

Returning home, Aidan made contact with various people including officials of the WJCB and eventually a committee was formed in 1999, known as "The Year of the Jersey Association". Its task was to organise and raise funds to create a permanent tribute to the Jersey cow in its Island home in the form of a bronze public sculpture.

During the ensuing two years, the 'association' worked extremely hard to accomplish its target of fund-raising, which came from a generous public response from every sector of Island life as well as overseas Jersey breeders and organisations.

The States of Jersey formally recognised 2001 as "The Year of the Jersey", as a tribute to the Island breed and a commemoration of the 50th

• John McKenna, the sculptor, poses with one of the bronze cows.

• Part of the large crowd at the unveiling of the sculpture in West Centre.

anniversary of the World Jersey Cattle Bureau.

The association commissioned an English sculptor, John McKenna, a Member of the Royal Society of British Sculptors, to create a life-size bronze group of Jersey cattle. John produced a magnificent sculpture, leaving Jersey with an enduring tribute to the Jersey cow, its breeders in the Island and worldwide.

Sir Philip Bailhache, Bailiff of Jersey, and Patron of the Year of the Jersey Association unveiled the sculpture on 29th September 2001.

- The front view of the sculpture.

- Looking at the sculpture from the side, showing part of Bath Street.

• Going home

bibliography

Ahier, Philip, M.A., B.Sc. and Ashworth, W S, A Short Parochial and Commercial History of Jersey, (Ashton & Denton Publishing Co. (C.I. Ltd)

American Jersey Cattle Club, The Jersey Review, Number Three, (The American Jersey Cattle Club)

Blackstone, Chris and Quesne, Katie Le, ST MARTIN, JERSEY, the Story of an Island Parish (published by Phillimore & Co. Ltd, 1999)

Boston, Eric J, Jersey Cattle, (Faber and Faber Limited 1954)

Brocq, Edward Le, The Island Cow, Volume 1, Nos 1 to 8, Jersey Channel Islands.

Coutanche, Henry, Coutanche, Sheila and Feuvre, Suzanne Le, St Lawrence Jersey, (the Parish of St. Lawrence, 1999)

Dain, John Le, Jersey Alphabet, (Seaflower Books 1997)

Department of Agriculture and Fisheries, Annual Statistics of the

Feuvre, David Le, Jersey: Not Quite British, (Seaflower Books 1993)

Gow, R M, THE JERSEY, an Outline of her History During Two Centuries –1734 to 1935, (The American Jersey Cattle Club New York, N.Y. 1938)

Hillsdon, Sonia, Jersey's Farmers and Their Fields, Book Six, (Sonia Hillsdon, 1994)

Jenkins, Harry, Famous Jersey Cattle, Volume 1, (Harry Jenkins & Son, Downers Grove, 1922)

Jersey Cattle Breeders' Society of South Africa, The South African Jersey Herd Book, Volume 1, BULLS 1-385, COWS 1-690, (The South African Stud Book Association 1928)

Jersey Island Federation of Women's Institutes, JERSEYS, OLD and NEW, The "Traditional" Jersey (Jersey Island Federation of Women's Institutes 1970)

Lemprière, Raoul, History of the Channel Islands, (Robert Hale & Company, Publishers 1974)

Maistre, Frank Le, Dictionnaire Jersiais-Francais avec Vocabulaire Français-Jersiais par Albert L. Carré, (Don Balleine Trust, 2 Hill Street, Jersey 1966)

Manning, John, The Parish Pump, a guide to discovering the real Jersey, (Channel Television Publications, T V Centre, Jersey 1987)

Mollet, Ralph, A Chronology of Jersey, Fourth Editio,n (Société Jersiaise 1949)

Nel, Dr G D, JERSEYS in Southern Africa, (Tafelberg, Cape Town 1968)

Nuttall, G T, The Jersey Breed, History of its Development and Progress in Queensland, (The Jersey Cattle Society of Queensland, Australia)

Perchard, Anne, The Jersey Cow, (Société Jersiaise 1999)

Perrée, compiled by J A, Notable Jersey Cattle, (Albert Smith, Photographer, Jersey)

Planning and Environment Committee, Roadside Walls, Fossés and Banques, Repairing common faults (Planning and Environment Committee)

Reference Library, Highlands College and Philip Mourant Training Centre

RJA & HS, A Pageant of Country Life Souvenir Programme 1833-1983

Royal Jersey Agricultural and Horticultural Society (Agricultural Department) Annual Reports of,

Royal Jersey Agricultural and Horticultural Society (Agricultural Department), Minutes of,

Royal Jersey Agricultural and Horticultural Society, RJAS 150th Anniversary Brochure, (Walker's Advertising Associates, Jersey 1983)

Shepard, H G, One Hundred Years of the Royal Jersey Agricultural and Horticultural Society, 1833-1933, (RJA & HS, 1934)

Société Jersiaise, Société Jersiaise Annual Bulletin for 1989, Volume 25, (Société Jersiaise, 1989)

Stevens, Joan, A Short History of Jersey, (La Société Jersiaise)

Stratton, J M and Houghton Brown, Jack, Agricultural Records A.D. 220-1977, (John Baker Publishers Limited, London, 1978)

Sueur, Frances Le, Flora Of Jersey, (the Société Jersiaise, Jersey 1984)

Syvret , Marguerite and Stevens, Joan, Balleine's History of Jersey (Phillimore & Co Ltd, 1981)

Tapes and references from the Jersey Archive Service

Thornton, John, The English Herd Book of Cattle, Vol II BULLS and COWS 974-1294, (A Committee of Jersey Breeders of Great Britain, 1880)

Thornton, John, The English Herd Book of Jersey Cattle, Vol 1BULLS 1-973, (A Committee of Jersey Breeders of Great Britain, 1879)

Tubb, L Gordon, M A, The Book of the Jersey, (W E Baxter Limited)

World Jersey Cattle Bureau, Minutes of,

ROYAL JERSEY AGRICULTURAL & HORTICULTURAL SOCIETY EST 1833